Explaining the Weather

Student Exercises and Teacher Guide for

Grade Ten Academic Science

Jim Ross   *The University of Western Ontario*

Mike Lattner   *Algonquin and Lakeshore Catholic District School Board*

 *London, Ontario Canada*

National Library of Canada Cataloguing in Publication	Ross, Jim (James William), 1952- Explaining the weather : student exercises and teacher's guide for grade ten academic science / Jim Ross. ISBN 978-1-897007-15-0 1. Weather--Study and teaching (Secondary)--Activity programs. 2. Meteorology--Study and teaching (Secondary)--Activity programs. 3. Weather--Study and teaching (Secondary) 4. Meteorology--Study and teaching (Secondary) I. Title. QC863.R68 2004 551.5'071'2 C2004-902579-1
Authors	Jim Ross Mike Lattner
Contributors	Lise Hoogkamp, Environment Canada Bob Cluett, USCG (ret.)
Printer Cover Design	CreateSpace Images, London, Ontario Canada

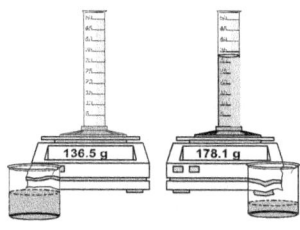

© Copyright 2003 by Ross Lattner Educational Consultants.

All rights reserved. The use of any part of this publication, reproduced, transmitted in any form or by any means, electronic, mechanical, photocopying, recording or otherwise, or stored in a retrieval system, without the prior consent of the publisher, is an infringement of the copyright law and is forbidden.

Permission is granted to the individual teacher who purchases one copy of *Explaining the Weather*, to reproduce the student activities for use in his / her classroom only. Reproduction of these materials for an entire school, or for a school system or for other colleagues or for commercial sale is strictly prohibited.

ISBN	978-1-897007-15-0
Offices	London Ontario Canada

To teachers, parents and students everywhere who desire to bring about new ways of understanding the world.

We welcome your comments and suggestions. Let us know what you find most useful. We've worked hard to remove any errors, but don't let a day go by without letting us know if you find one.

Stay in touch.

Our thanks to all of the wonderful people at the Faculty of Education, the University of Western Ontario and St. Paul Catholic Secondary School.

Special thanks to Lise Hoogkamp, Environment Canada, and Bob Cluett, USCG (ret.), for technical advice.

Explaining the Weather
Table of Contents

Teaching About the Weather 1
   Unit Planning Notes 2
   Activity 1.1: Your Daily Weather Log 4
   Activity 1.2: Conduction, Convection and Radiation 6
   Activity 1.3: Energy and Changes of State of Water 8
   Lab 1.4: The Concept of Atmospheric Pressure 10
   Activity 1.5: Atmospheric Pressure 12
   Activity 1.6: Temperature in the Atmosphere 12
   Lab 1.7: Water Vapour, Humidity and the Dew Point 14
   Activity 1.8: Relative Density of an Air Parcel 16
   Lab 1.9: Formation of a Cloud 18
   Activity 1.10: How Far Will an Air Parcel Rise? 20
   Quiz 1.11: Energy and the Weather 22
   Lab 2.1: The Concept of an "Air Cell" 24
   Activity 2.2: Air Cells, and the Formation of Cumulus Clouds 26
   Activity 2.3: The Chinook 28
   Activity 2.4: Air Masses and Atmospheric Pressure 30
   Activity 2.5: Warm Fronts 32
   Activity 2.6: Cold Fronts 34
   Activity 2.7: Atmospheric Pressure and Isobars 36
   Quiz 2.8: Air Circulation 38
   Activity 3.1: We Live on a Spinning Ball 40
   Activity 3.2: Hadley Cells and Prevailing Winds 42
   Activity 3.3: The World in January 44
   Activity 3.4: The World in July 44
   Activity 3.5: Pressure Systems and Weather Maps 46
   Activity 3.6: Summer Front Systems in Southern Ontario 48
   Quiz 3.7: Spinning the Weather 50
   Project 4.1: Extreme Weather Events 52

Explaining the Weather
Table of Contents

Explaining the Weather 55
    Introduction: Three Theories to Understand Our Weather............................... 56
    Activity 1.1: Your Daily Weather Log.. 58
    Activity 1.2: Conduction, Convection and Radiation..................................... 60
    Activity 1.3: Energy and Changes of State of Water 62
    Lab 1.4: The Concept of Atmospheric Pressure... 64
    Activity 1.5: Atmospheric Pressure 66
    Activity 1.6: Temperature in the Atmosphere 68
    Lab 1.7: Water Vapour, Humidity and the Dew Point 70
    Activity 1.8: Relative Density of an Air Parcel 72
    Lab 1.9: Formation of a Cloud 74
    Activity 1.10: How Far Will an Air Parcel Rise?... 76
    Quiz 1.11: Energy and the Weather 78
    Lab 2.1: The Concept of an "Air Cell"... 82
    Activity 2.2: Air Cells, and the Formation of Cumulus Clouds 84
    Activity 2.3: The Chinook 86
    Activity 2.4: Air Masses and Atmospheric Pressure 88
    Activity 2.5: Warm Fronts 90
    Activity 2.6: Cold Fronts 92
    Activity 2.7: Atmospheric Pressure and Isobars... 94
    Quiz 2.8: Air Circulation 96
    Activity 3.1: We Live on a Spinning Ball 100
    Activity 3.2: Hadley Cells and Prevailing Winds 102
    Activity 3.3: The World in January 104
    Activity 3.4: The World in July 106
    Activity 3.5: Pressure Systems and Weather Maps 108
    Activity 3.6: Summer Front Systems in Southern Ontario........................... 110
    Quiz 3.7: Spinning the Weather 112
    Project 4.1: Extreme Weather Events 116

Appendix: Laboratory Safety 118

Explaining the Weather
Teaching About the Weather

Title:	Explaining the Weather
Time Allocation:	27.5 hours (22 periods of 75 minutes each)
Authors:	Jim Ross and Mike Lattner
Date:	May 2003

Unit Description:

This unit provides students with the tools to analyze the first order forces that influence Earth's weather and is subdivided into three major sections. Each section will take a little more than one week to complete.

1. Basic structure of the atmosphere and the factors that shape it. Foremost is the particle nature of the gases that make up the atmosphere, and the forces that act upon the gases. Also included is the role of water and its changes of state in the energy economy of the atmosphere.

2. The circulation of air parcels as they absorb and release energy from the sun. This fundamental motion is simple, but can be used to explain complex phenomena like warm and cold fronts.

3. The effect of the spinning of Earth upon the motion of circulating air masses. The winds that we experience are shaped more by the Coriolis effect than by thermal effects. This counter-intuitive idea is one of the more difficult ideas for a child to learn.

At the end of the formal exercises, students are invited to undertake a five day research project on a weather event of interest to them¢.

Strand:
Earth and Space Science

Expectations:
Overall Expectations: ESV 1 - 3
Specific Expectations: ES1.01-1.07, 2.01-2.03; 3.01-3.03

Explaining the Weather

The weather is experienced by children in two predominant ways: direct experience, and mediated experience (newspapers, television and the internet).

Direct experience is so vivid and so varied, that it is very difficult for a student to "take it apart." Thus, weather almost defies explanation.

Mediated experience is simple in its reductionism, yet so obscure in its origins, that most students find they cannot use mediated accounts to construct anything like a genuine explanation.

If good *science* is about explanation, then we must find a way to introduce students to a set of representations that allow them to really explain the weather in terms of simpler, more permanent things.

Unit Planning Notes
This unit is designed to provide students with a set of simple representations of the weather which permit a student to explain in some detail the processes that underlie observable events in the weather around them.

Prior Knowledge Required This unit assumes that the student has a sound grasp of the particle theory, and understands the basic forces of gravity and electricity. While a typical student does not have a strong, theory-based understanding of the objects in the universe, we assume that he or she does have some degree of commitment to a coherent scientific world view.

Teaching and Learning Strategies Science is not the study of nature itself. Science is the study of our shared *representations* of nature. Accordingly, three learning strategies are emphasized.

1. Students are expected to commit themselves to a *prediction* of the behaviour of each demonstration or lab.

2. Students are expected to explain why they believe their prediction, in both *pictorial representations* and in *sentences*.

3. Students are expected to gradually master a small set of *theoretical propositions*, and then to increasingly represent their arguments in terms of the theory.

For example, the question *"Why is it raining today?"* could be answered in different ways. Here are three examples:

"This is the rainy season. It always rains at this time of year."
"A cold front has pushed in from the northwest."
"This is nature's way of reducing the humidity."

Each of these "answers" corresponds to a particular purpose, that is not necessarily scientific. We want our students to be able to relate events in the weather to other scientific systems of explanation, such as the kinetic molecular theory of matter, conservation of momentum and energy, and so on.

Assessment and Evaluation A variety of strategies and instruments will be used throughout this book. A final project is a major component in this unit. Use the KICA wheel to identify for students the area of learning that you wish to emphasize in each exercise.

Introduction

Science and Pedagogy

The student learning goals of this unit are:

To explain common weather phenomena in terms of the particle theory.

To explain the role of the changes of state of water in the energy of weather systems.

To explain the movement of air masses in terms of circulation of air parcels and the rotation of Earth.

Accomplishing these goals requires that students gradually learn to represent weather in terms of particles, energy, and moving air parcels.

This unit consists of three main theoretical ideas:

1. **The Particle Theory of Matter** consists of six simple statements which can help you explain things that happen all around you, and even help you to predict things you have never seen!
   1. The absence of matter is a pure vacuum.
   2. All matter is made of tiny particles.
   3. All particles of one substance are identical.
   4. The spaces between particles are small in solids and in liquids, and large in gases.
   5. All particles are attracted to each other by forces.
   6. Particles are in constant motion.

2. **The Concept of an Air Parcel**
   1. The *mass* of an air parcel is equal to the total mass of its particles.
   2. The *temperature* of an air parcel is a measure of the speed of the molecules.
   3. The *pressure* of an air parcel depends upon the masses of the parcels above it.
   4. The *volume* of a parcel gets larger with increased temperature.
   5. The *volume* of a parcel gets smaller as air pressure is increased.
   6. The *relative density* of an air parcel is equal to its *mass ÷ volume* (D = m/V).

3. **A Theory of the Atmosphere**
   1. All air molecules have mass, so they experience a downward force of gravity.
   2. All of the air molecules above a parcel of air press downward on that parcel.
   3. Kinetic energy from sunlight makes molecules move at about 450 m/s near the ground.
   4. Each parcel of air supports all of the molecules above it.
   5. Warmer (faster) particles expand a parcel; cooler (slower) particles are squeezed together.
   6. At the same pressure, cool air is more dense than warm air.

Several other representations will also be required of the students, including maps.

Explaining the Weather

10 Academic Science Teachers' Guide

Activity 1.1: Your Daily Weather Log

Pedagogical Issues We are calling upon students to study the weather around them. Our first task is to draw the student's attention to the weather. One way to accomplish this is to make the weather into a problem for students to solve. But how do we make the act of paying attention to the weather problematic? One way is to make commercial weather predictions problematic. How accurate are weather predictions published in our media?

In this activity, then, students must begin to use the meteorologists' language, and also to compare the meteorologists' predictions to the students' observations. The problem is the existence (or even the anticipation of) discrepancies between the professional predictions and the observed weather.

The language of professional meteorologists contains a large number of concepts which seem to appear nowhere else in the students' experience. This creates an obstacle for students learning. For example, the concept of high pressure would seem utterly disconnected from a student's typical everyday experience of pressure. We don't feel any change in pressure on our skin.

Science Issues In the scientific community, what passes as a prediction? Are the nightly prognostications of the TV weather person equivalent to a scientific prediction? Does the TV weather person even provide good explanations? We may be told that the reason it is cold and wet outside is because the jet stream has meandered south. Does that kind of thing count as a coherent explanation of the weather outside?

These are not idle questions. The student's understanding of the nature of science is at stake here. If we wish our students to understand science as a coherent system of understanding and explaining the world, our standards of explanation in this unit must be high.

In spite of the huge impact of media, the students who come to us are often impoverished in personal experiences.

Students tend to experience pressure as the cause of things. For example, if you wish to cool your soup, you increase the pressure in your lungs (you can feel it!) and you blow upon the soup. Pressure appears to cause changes.

Is pressure the cause of the weather, or just an effect of other processes? Does, for example, high pressure "cause" blue skies? Or does the prevalence of blue skies cause high pressure?

Energy and the Weather

Science and Pedagogy

The Learning Activity

Each student, or pair of students, must select one weather source, either a television broadcast, an internet source, or a print media source. Ideally this can be accessed at home.

Each evening, students are asked to look specifically at the 3-day forecasts for the local weather (not the general forecast, and not tomorrow's forecast). The students are to record each forecast for four weeks. Each day, the students are to observe the weather and compare the weather to the forecast made three days earlier.

Before the activity, students will

An appropriate table for this activity will have four columns:

- Date prediction made
- Prediction
- Date of observed weather
- Observed weather

Predict: what percentage of the weather forecaster's predictions will be correct? 90%? 50%? The student is to commit to a percentage.

Explain: this section is an attempt to elicit the student's ideas of how forecasters make predictions.

During the next four weeks, students will

Observe: one weather source for 4 weeks and the weather itself for 4 weeks, and one satellite weather map each day (*NB. You may wish to post only one map for the whole class*).

After the experiment, students will calculate the % accuracy over the four weeks. Finally, students will

Explain: any differences between what the student predicted in meteorologist accuracy and observed accuracy.

Equipment, Preparation and Resources
- a variety of print and electronic weather sources, (students can also access these at home)
- large paper for data tables

Categories: | **Assessment and Evaluation**
Knowledge and Understanding:
Thinking and Inquiry: | Frequency, consistency and quality of the data collected
Communication: | Clarity and depth of the presentation and questions
Applications / Connections:

10 Academic Science Teachers' Guide
Explaining the Weather

Activity 1.2: Conduction, Convection and Radiation

Pedagogical Issues

The experience of cause and effect are very much located in the body. Upon seeing an "effect", a student can hardly refrain from identifying a "cause". This action may have very deep roots in our nervous systems. However, part of the discipline of science is to withhold judgements of cause until a large number of contributing factors have been identified.

In this case, it is very simple for a student to say "Hot air rises" and stop thinking. Repeated references to the particle theory of matter and to changes in the motion of the particles challenges the student to think about more factors than simply the temperature.

Science Issues

Solar energy reaching the upper atmosphere is arriving at a rate of 1367 J/m²•s. In one hour, 3600 s, the total solar energy arriving in the upper atmosphere is 4.92 MJ.

Half of that energy, 2.46 MJ, is converted to heat in the black asphalt pavement.

The temperature change of the pavement, ΔT, can be calculated:

$$\Delta T = \frac{Q}{mc}$$

$$= \frac{(2.46 \times 10^6 \text{ J})}{(500 \text{ kg})(900 \text{ J/kg}°C)}$$

$$= 5.5 \text{ °C}$$

A similar calculation for 10 000 kg of air yields a much smaller temperature change from direct heating of the atmosphere (0.15 °C).

Clearly, the earth's atmosphere is not heated directly by the sun.

A cat sees the grass move, and begins to pay close attention to the movement. Is the cat making some kind of inference that the grass is being moved by a small animal, and not by the wind?

The same cat watches the grass moving, and leaps to the place where the mouse is likely to be. Is the cat making some kind of prediction about the future location of the mouse?

Obviously, the cat is doing something that a snail cannot do. Could you call the cat's activities inference and prediction? If so, in what ways are human inference and prediction biologically programmed into us?

Energy and the Weather

Science and Pedagogy

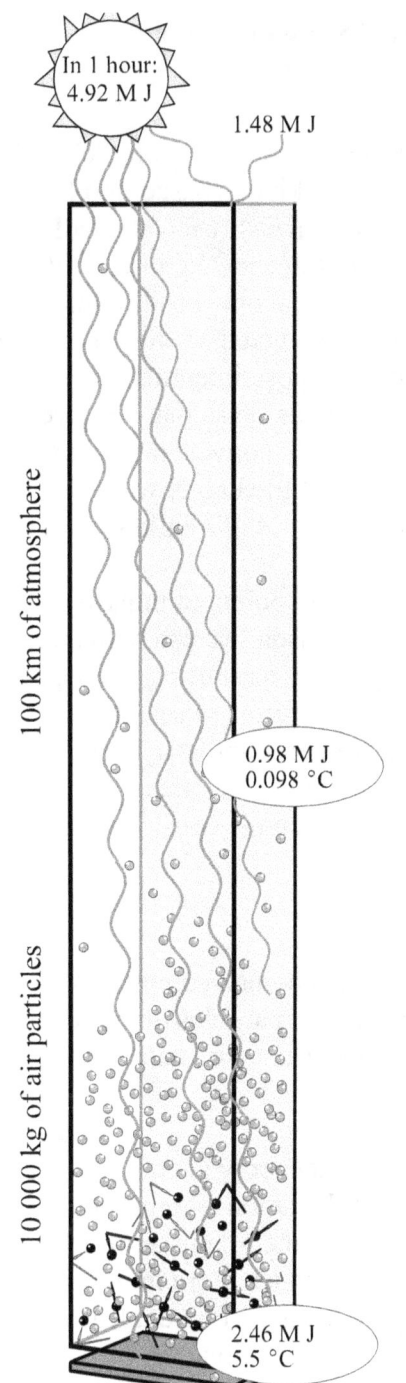

The Learning Activity

Primarily a pencil and paper exercise in representing energy gains and losses in the atmosphere, this exercise will require considerable teacher guidance to get most students started.

Review the process of rearranging algebraic expressions and ensure that students include units in each calculation.

Equipment, Preparation and Resources

This activity is a pencil and paper exercise.

A qualitative demonstration may support the conclusions. A small low voltage spot light directed at a blackened block of metal will increase noticeably in temperature, but the air between the lamp and the block will not.

Categories:	Assessment and Evaluation
Knowledge and Understanding:	Describes general conclusions clearly
Thinking and Inquiry:	Arrives at supportable conclusions by completing math correctly
Communication:	Clarity and completeness of work
Applications / Connections:	

© Ross Lattner Publishing   www.rosslattner.ca

Explaining the Weather

10 Academic Science Teachers' Guide

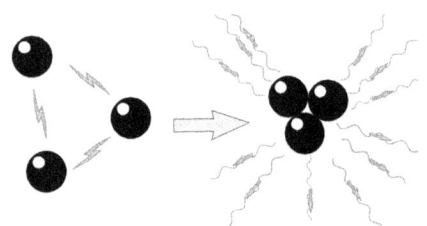

Whenever particles are attracted together, they give off energy as they slam together. Think of the heat and noise given off when two magnets click together.

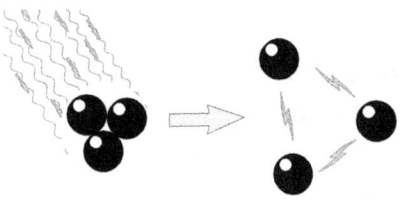

Energy must be added to break a clump of attracted particles apart. Think of work you must do to pull two magnets apart.

Seldom does one read that "*a coherent understanding of science*" is a stated goal of science curriculum documents.

Activity 1.3: Energy and Changes of State of Water

Pedagogical Issues

Many thinkers suggest that the human mind is built upon the human body in such a way that human thinking is based upon human bodily actions. For example, we can easily move small objects around with our arms, hands and fingers. We intend that little ball X is to be picked up and moved to a new position in order to accomplish a purpose that we have constructed. Some of our thinking processes reflect the structure of those intentional movements. At the same time, the conscious sense that we give to our mental and physical actions is frequently conveyed in a story-like structure. When a child explains *why* little ball X was moved, the explanation has many of the characteristics of a little story.

These two natural structures are the starting points of thinking. When we wish to teach children how to reason, we can help them by using these two deeply ingrained patterns as starting points, and gradually converting them to more general forms of scientific reasoning. By incorporating body movements and story structures into our instruction, movements and structures that are already understood by the student, we can build upon the considerable natural intelligence of our students.

Science Issues

In the study of weather, the underlying phenomena are the movements, aggregations, and dispersion of particles of matter. In particular, the energy associated with the change of state of water has enormous effects upon the weather around us.

A student's ability to simply calculate that energy is not a demonstration of understanding. The calculation is important, to be sure, but the student's ability to explain melting in terms of particles, forces, movement and energy is much more important. Our goal should always be meaningful, lasting, coherent learning. That goal is much bigger than the calculations alone.

The particle pictures challenge students to depict the solid, liquid and gas states many times. There should be the same number of particles in the two state diagrams (10 - 12 is usually enough). This sort of representation can provide many opportunities for students to learn to "see" what is going on when, for example, water vapour turns to ice.

Note that the attractive forces between the water particles are electric forces, (i.e., van der Waals forces, dipole-dipole interactions, and hydrogen bonds). The sun's radiant energy is electromagnetic energy. Photons are, of course, the carriers of the electric force between charged particles. Particles exchange energy by exchanging low energy photons, such as infrared.

Energy and the Weather

Science and Pedagogy

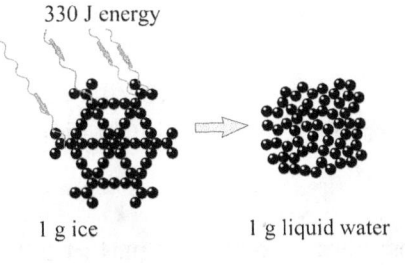
330 J energy
1 g ice → 1 g liquid water

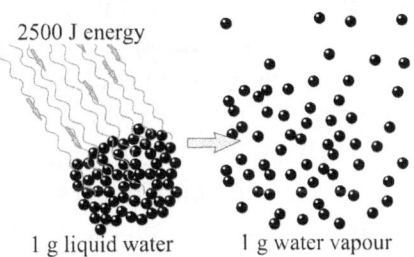
2500 J energy
1 g liquid water → 1 g water vapour

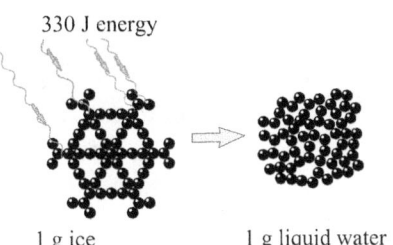
330 J energy
1 g ice → 1 g liquid water

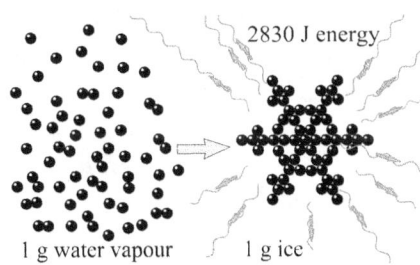
2830 J energy
1 g water vapour → 1 g ice

If 25 grams of water vapour turned to ice, how much energy would be give off?

Particles of water are constantly "communicating" with other particles via photons of light. When the particles move together, there is a net radiation of energy outward into the environment, heating it. When particles of water move apart, there is a net absorption of energy from the environment, cooling it.

The Learning Activity Both the pictures and the words are important in these exercises. In order to attain the ability to explain in a coherent fashion such diverse phenomena as the chinook and the killing frost, one must be able to represent both phenomena in terms of something both simpler and more permanent. The particles of matter are the "simpler and more permanent" things that underlie both events.

Equipment, Preparation and Resources
- worksheets provided in the student exercises
- pencils, pens calculators, etc.

Answers to student problems

1   a   To melt 1.0 kg of ice requires 330 kJ
     b   1000 J could melt 3.0 g of ice
     c   2.46 MJ could melt 7.5 kg of ice, making 7.5 L water

2   a   625 kJ is required to dry up 250 g of water
     b   144 g of water could be vapourized
     c   The sun could dry up 984 g water in 1 h (0.98 L)

3   The sun could dry up 869 g of snow in 1 h

4   That patch of L. Eire would release 6.6×10^{14} J into the atmosphere, enough to keep fruit orchards from freezing

5   a   30 000 J would be released
     b   It would take only 4 g of water to condense
     c   2.5 MJ would be released by the dew

6   8.5 million joules was released when the snow formed.

Categories:
Knowledge and Understanding:
Thinking and Inquiry:
Communication:
Applications / Connections:

Assessment and Evaluation
Accuracy and consistency of explanations

Clarity of explanations, diagrams

© Ross Lattner Publishing   www.rosslattner.ca

Explaining the Weather

10 Academic Science Teachers' Guide

Lab 1.4: The Concept of Atmospheric Pressure

Learning Expectations ES1.01: Identify and describe the principal characteristics of the hydrosphere and the four regions of the atmosphere.

Pedagogical Issues The atmosphere is so much a part of our environment that we seldom think of it. For a student, the concept of atmospheric pressure is largely unconnected with everyday experience. The absence of any experiential connections creates a difficulty for the student. If the student is used to accepting the teacher's authority, the teacher need only assert that the atmospheric pressure is 101.3 kPa, and the student will agree. But if the student needs to relate learning to personal experience and personal meaning, such an assertion on the part of the teacher is unlikely to mean much to the student. How can we set up an activity for students to experience atmospheric pressure?

Two components are part of this (and every) lesson: an experiential component, and a representational component.

The experiential component is essential to provide a point of reference for the student. Experience alone is not sufficient to support further reasoning, however.

The representational component is necessary to provide a coherent platform from which to reason about pressure.

Science Issues

We live at the bottom of an ocean of air. This ocean exerts considerable pressure upon us, but we are usually unaware of it. The pressure itself is due to the force of gravity on the atoms of the atmosphere. Since most of the mass of an atom is located in its nucleus, we might say that atmospheric pressure is the force of gravity upon the heavy nuclear particles in the column of air above us.

Atmospheric pressure is so large that even small variations in the number of heavy particles overhead can still have considerable influence on the behaviour of air currents down on the surface of the earth. Consider a difference in pressure of only 10 mb (1005 mb to 995 mb). In the column of air above the 1m × 1m area in which you are sitting, this represents a difference of 100 kg of heavy particles! Suppose city block B (500 m × 500 m) had pressure 995 mb and the adjacent city block A had pressure 1005 mb. That would mean that city block A has 25 000 000 kg of additional air particles overhead. This would result in very high winds, hundreds of kilometres per hour. Obviously, pressure differences are much smaller and more spread out than our example.

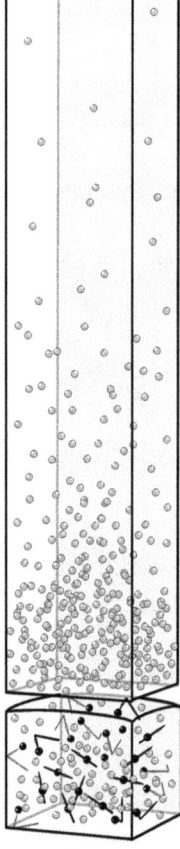

Energy and the Weather

Science and Pedagogy

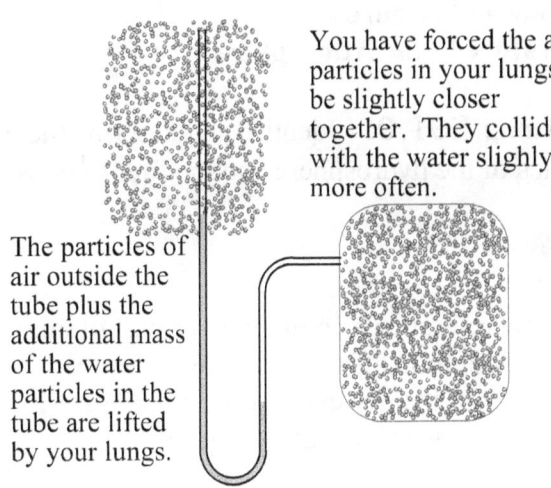

The particles of air outside the tube plus the additional mass of the water particles in the tube are lifted by your lungs.

You have forced the air particles in your lungs to be slightly closer together. They collide with the water slighly more often.

1.29 m of iron

exerts the same pressure as

100 km of air !

Why? Both columns contain the same number of heavy particles (protons and neutrons) They are much more closely packed in the iron.

Learning Activity One
Before the experiment, students will:
Predict: How high can you lift a column of water, using air pressure developed by your lungs?
Explain: why they believe the prediction, and
Observe: how high they can push the water.

After the experiment, students
Explain: any discrepancies between the prediction and observation, using both pictures and sentences.

The column of air over your head is 100 km high, and has the same number of heavy particles (protons and neutrons) as a column of water only 10 m tall. To exert one atmospheric pressure, your lungs would have to lift the column of water to a height of 10 m! How far were students from that goal?

Learning Activity Two:
What is the mass of the air above us? Prepare a steel bar (steel is almost pure iron) exactly 1.29 m in height. The mass of the nuclei in the steel bar is equal to the mass of the nuclei in a similar column of air from earth to space. Believe it or not, the pressure of the steel bar on your hand is the same as the pressure of the atmosphere!

Equipment, Preparation and Resources
- one steel bar, any diameter, exactly 1.29 m in height
- a manometer, constructed of plastic tubing fastened to a plywood board. The total vertical height of the manometer is 1.0 m, with the mouthpiece at 0.60 m. You must supply drinking straws for sanitary mouthpieces. Add food colour to the water and fill the manometer to a height of 0.5 m.

Categories: **Assessment and Evaluation**
Knowledge and Understanding: Calculates the pressure of a column of water
Thinking and Inquiry:
Communication:
Applications / Connections: Relates atmospheric pressure to other pressures

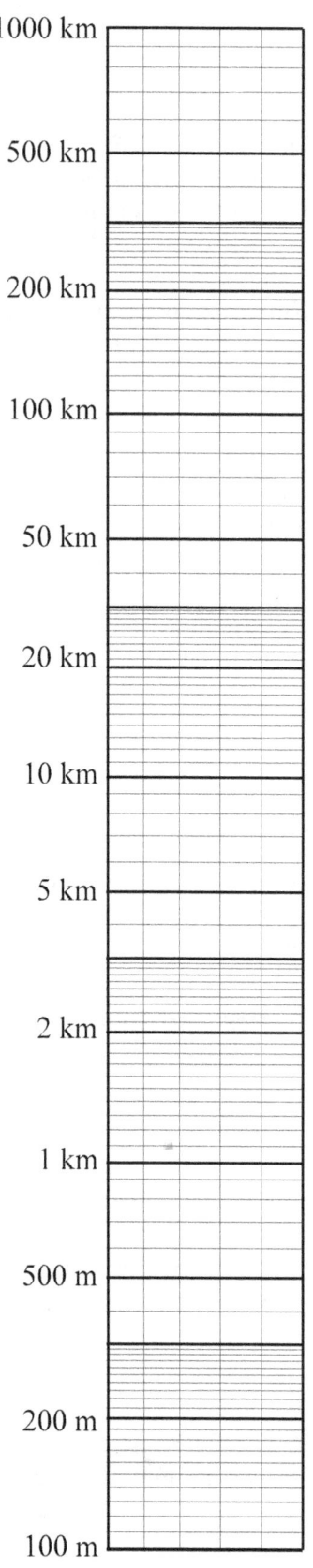

Activity 1.5: Atmospheric Pressure
Activity 1.6: Temperature in the Atmosphere

Learning Expectations ES1.01: Identify and describe the principal characteristics of the hydrosphere and the four regions of the atmosphere.

Pedagogical Issues

Once again, the atmosphere is such a commonplace part of our environment that we seldom think of it. Furthermore, even though we see the clouds and the sky every day, we cannot perceive directly how far away or how large they are.

It is the purpose of this exercise to have students link their personal experiences to a system of representation, so as to make as many personally meaningful connections as possible. For example, almost all students have observed passenger jets at high altitudes, and many of your students have traveled in one. A student who can relate those experiences to the temperature and pressure conditions that prevail outside such a jet is more likely to remember the new information.

In addition, we expect students to learn how to use a large number of terms in this unit. Without the context provided by meaningful personal experience, such terms as troposphere, stratosphere, mesosphere, thermosphere, ionosphere and ozone layer are unlikely to be remembered. It is pedagogically unsound to attempt to memorize such definitions in the absence of a meaningful context.

Science Issues

This may be the first time that your students have encountered semi-logarithmic graphs. There is no need to make a big deal of it. Point out that the scale of such a graph allows us to describe a large range of altitudes on a single sheet of paper.

To a student who is ready, you might point out that no matter where you start on the graph paper in the student exercises, moving your pencil 53 mm upward corresponds to a tenfold increase in altitude. Moving your pencil upward 35 mm corresponds to an altitude 5 times higher. How many millimetres corresponds to doubling the altitude?

Energy and the Weather

Science and Pedagogy

What other items could students add to the graphs?

Landmarks
Mt Everest
Aviation Altitude Records
Birds
Insects
Bacteria
Spores
Pollution
Disasters

The Learning Activity

Students work together to place various objects and measurements on the graph paper.

Emphasize to students that the atmospheric pressure decreases steadily from the ground up, whereas he temperature graph, on the other hand, undergoes large variations as one ascends.

The temperature boundaries define the various "spheres". It may actually be more correct to say that energy - related phenomena such as the ionization of air molecules, the formation of ozone, and the dissociation of oxygen into atoms define the boundaries of the layers of the atmosphere.

Equipment, Preparation and Resources

Students will use the work sheets from the student exercises, pencils, and rulers.

Categories:
Knowledge and Understanding:
Thinking and Inquiry:
Communication:
Applications / Connections:

Assessment and Evaluation

Correctly identifies the trends in pressure and temperature
Generates further questions about pressure, temperature and heat

Explaining the Weather

10 Academic Science Teachers' Guide

> Linear thinking: one clear causative step at a time. This is typical of "natural thinking" and can be observed in very many settings.
>
> Young people are usually quite confident at thinking in this mode as long as the cause and the effect relationship is personally meaningful.
>
> Network thinking: competing influences generate a complex set of outcomes. No clear causative steps. This kind of thinking is not "natural." We observe it only in communities of particular disciplines.
>
> Human beings of all ages tend to avoid "network thinking" when faced with a network phenomenon. Instead, we approach the phenomenon with "natural thinking." We try to explain the situation one "linear thinking" step at a time.
>
> Now, we teachers can either "go with the flow," or "swim against the current."

Lab 1.7: Water Vapour, Humidity and the Dew Point

Learning Expectations 1.02: Describe and explain heat transfer within the water cycle and how the hydrosphere and atmosphere act as heat sinks.

Pedagogical Issues

Two distinct pedagogical concerns arise here. First, there are many ways to describe the related phenomena of water vapour, dew, the dew point, and relative humidity. Some ways of defining these terms make them inaccessible to students. Second, students tend to think linearly, that is, one clear causative step at a time. Network thinking, in which competing influences generate a complex set of outcomes, is not natural to students. Network thinking requires years of attention, in any discipline. We can address these two issues simultaneously if we carefully choose the units of analysis that we teach.

We chose to define a parcel of air as exactly 1.00 kg of dry air. This choice provides us with a simple method of determining what mass of water is dissolved in a parcel of air at any given temperature.

Science Issues

We may profitably use the concept of a solution here. Air is a solution. If we consider water to be the solute, then dry air is the solvent.

Let us consider a parcel of air at 26 °C.

a) If the air parcel was saturated, what mass of water vapour could it hold?

b) If the parcel was at 60% relative humidity (RH), what mass of water could it hold?

c) If the parcel was cooled to 6 °C before dew began to form, then what mass of water did it actually hold?

d) Given (a) and (c), what was the actual % humidity of the air?

Energy and the Weather

Science and Pedagogy

T °C	grams H₂O
40	48.4
38	43.5
36	38.8
34	34.5
32	30.7
30	27.2
29	25.6
28	24.1
27	22.7
26	21.3
25	20.1
24	18.9
23	17.8
22	16.7
21	15.7
20	14.7
19	13.8
18	13
17	12.1
16	11.3
15	10.7
14	10
13	9.3
12	8.7
11	8.1
10	7.6
9	7.1
8	6.6
7	6.2
6	5.8
5	5.4
4	5.0
3	4.7
2	4.4
1	4.0
0	3.8
-1	3.5
-2	3.3
-3	3.0
-4	2.8
-5	2.5
-6	2.4
-8	2.1
-10	1.7
-12	1.5
-14	1.3
-16	0.9
-18	0.8
-20	0.6

The Learning Activity

Before the experiment, students will
- **Predict:** the % RH in the classroom, and
- **Explain:** their predictions.

At this point, students must make up an experiment to measure %RH using very simple equipment. You know your class. Your students may be able to design this experiment by themselves, or they may need some guidance. A typical experiment is to:

1. Prepare a can about half full of warm water, above room temperature.
2. Measure the temperature of the water, and record.
3. Add ice, and stir. Measure and record the temperature of the water at regular time intervals. At one particular temperature, dew begins to form on the outside of the can. That is the temperature at which air is saturated with water vapour.
4. Compare the *actual* mass of water in the air with the *maximum* mass of water the classroom air could have held at its original temperature.

After the experiment, students
- **Observe:** and record the %RH in the room then compare their measurement with other students.
- **Explain:** their measurement. Is it accurate? What could someone do to improve it?

Equipment, Preparation and Resources

To solve the problems, students use the printed pages in the student exercises.

For the experiment, each group of 2-4 students will need a thermometer, a metal can, water, ice, and paper towels.

Categories: Assessment and Evaluation

- **Knowledge / Understanding:** Uses the table correctly to calculate relative humidity *etc.*
- **Thinking / Inquiry:** Designs own experimental procedure
- **Communication:**
- **Application / Connection:** Can apply knowledge to a variety of other situations

An 1.00 kg air parcel at 0°C and 1013 mb has the volume shown below.

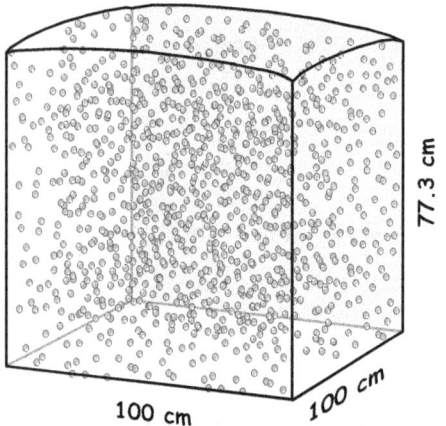

Warming the air parcel causes it to get larger, cooling causes it to shrink. A change of 1.0 °C makes the parcel 2.75 mm taller.

Water molecules (H_2O, FW = 18.0) are less massive than air molecules (O_2, FW = 32.0 and N_2, FW = 28.0). Adding 1.00 g of water vapour to an air parcel makes the parcel 0.50 mm taller.

Activity 1.8: Relative Density of an Air Parcel

Learning Expectations ES .05: Describe factors contributing to Earth's temperature gradients and wind speed and direction.

Pedagogical Issues

Density is a ratio, and hence is abstracted from the concrete experience of a student. Ratios, such as density, rates, and so on are more difficult for a student to learn than more directly experienced things. In this lesson, density is constantly related to the mass of the air parcel (chosen to be 1.00 kg of dry air) and the volume of the air parcel (which varies slightly with temperature and humidity). Indeed, the volume is further retracted back to a variation in the height of the air parcel, the area of the parcel being held constant at 1.00 m².

Science Issues

"Wait a minute!! Doesn't the air parcel follow the gas laws?" Are volume, temperature and mass not related to the equation:

$$PV = nRT$$

Yes, of course. But, how complex should the calculations be? Remember, students do not know the gas laws, nor the Kelvin temperature scale. Also, how precise must we be? In the present case, we must make the following assumptions in order for the computations to work as described.

First, the range of temperatures is restricted to -20°C to +40°C. Within that range, the 2.75 mm approximation is accurate within an average error of ± 0.5%. Second, it is true that adding 1.00 g water would increase both the mass and the volume of the air parcel. However, the overall effect of adding a gram of water upon the density is approximately the same as increasing the volume by 0.50 mm. Adding 5.0 g of water has about the same effect upon density as heating a dry parcel 1°C.

We are not comparing parcels at two different altitudes, so we don't need to include air pressure in our calculations. Overall, these approximate calculations will match the "correct" calculations to ± 1%. Obviously, this is not bad for our approximation.

Energy and the Weather

Science and Pedagogy

Consider two adjacent city blocks. Each block is 500 m × 500 m square. Over one of the blocks, the air is perfectly dry. Over the other block, the air is at 50% humidity up to an altitude of 5.0 km.

Each water molecule has only half the mass of an air molecule, but it occupies the same space.

The presence of the water vapour makes the total mass of the air about 40 kg less above each square metre, compared with the dry air. If you compare the city blocks, the moist air is about ten million kg lighter than the dry air.

When we are dealing with vast amounts of air such as exist in our atmosphere, even small changes have huge effects over large distances.

The Learning Activity

Each set of calculations is set up to result in a single prediction: will the air parcel sink, rise, or rest compared with the surrounding air?

Both temperature and humidity are included in the density calculations.

Equipment, Preparation and Resources

Pencils, paper, and calculators are required for the student exercises.

Categories:
Knowledge and Understanding:
Thinking and Inquiry:
Communication:
Applications / Connections:

Assessment and Evaluation

Understands effect of water content on density of an air parcel

10 Academic Science Teachers' Guide

Explaining the Weather

Lab 1.9: Formation of a Cloud

It is unlikely that a student can "discover" any meaningful connection between two things that lie outside everyday experience.

Learning Expectations ES1.03 & 1.04: Describe and explain heat transfer in the hydrosphere and atmosphere, and its effects on air and water currents; describe the effects of heat transfer within the atmosphere and on the development and movement of weather systems.

Pedagogical Issues We live at the bottom of an ocean of air, and the pressure that we experience is very nearly constant. It is very unlikely that we would have any direct experience of atmospheric cooling due to expansion of local air parcels.

Student: "Air cools when it expands?

That doesn't make sense!

We were taught that warm air expands.

Isn't that what the gas laws were all about? (PV = nRT)"

Likewise, when we or our students do ascend, in a small plane or on a trip to the top of a mountain, we certainly do experience our ears popping as the pressure decreases. We also experience the lower temperature at the higher altitude. Yet very few students or adults make any connection whatsoever between the reduction of atmospheric pressure, and the cooling of an air parcel. This connection is, like so many of the important ideas in this unit, outside the everyday experience of the student. This lab provides the students with one experience of the cooling effect of the expansion of air.

Science Issues Why does air become cool when it expands? Let's make a number of arguments that relate to more common experiences.

1. Imagine a stick of dynamite going off under a pile of bowling balls. Initially the balls are moving quite quickly. As they move upward *against the attractive force of gravity* they slow down. As they fall back down *with the force of gravity* they speed up. Now imagine the bowling balls are atoms: as they expand outward against an attractive force, they slow down (lower temperature). As they contract inward in response to an attractive force, they speed up (higher temperature).

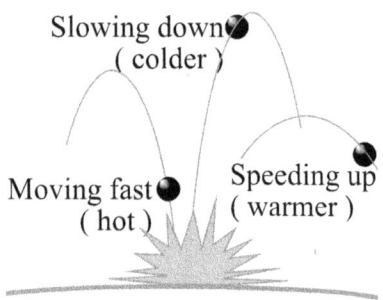

2. Imagine throwing a superball (atom) against a stationary wall. An ideal superball would bounce back at you with exactly the same speed (temperature, if we were talking about an atom). Now imagine throwing your superball at the front of an approaching locomotive. The superball would bounce back at you *faster* than you threw it. If a container was shrinking, an atom bouncing off the approaching walls of the container would be slightly hotter for the experience. Now imagine throwing your superball at the back of a receding train. The superball would bounce back at you slower than you threw it. An atom, bouncing off the walls of an expanding ...

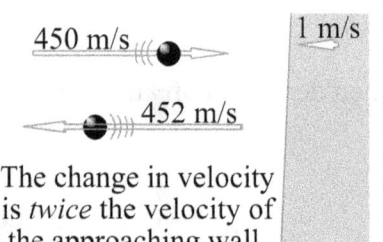

The change in velocity is *twice* the velocity of the approaching wall.

© Ross Lattner Publishing   18   www.rosslattner.ca

Energy and the Weather

Science and Pedagogy

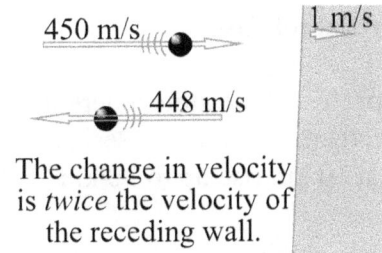

The change in velocity is *twice* the velocity of the receding wall.

The overall process is:

A warm, moist air parcel is less dense than the surrounding air parcels.

Gravity pulls the more dense parcels down, forcing the less dense parcel up.

The less dense parcel expands as it rises.

The expanding air parcel cools down.

When the cooling temperature reaches the dew point, moisture in the air parcel condenses on dust in the air.

A cloud is formed!

...container, would be slower (colder) for the experience. Particles of air traveling at 450 m/s could undergo millions of collisions in a very short time, so even if the wall of the container is moving slowly, the total ΔT could be quite considerable.

3. The conservation of energy argument. When you trap air in a plastic bag and squeeze it, you are doing work on the air in the bag. The work you do appears as thermal energy, or an increase in temperature. Likewise, when the bag is allowed to expand, the thermal energy inside the bag lifts the entire atmosphere just a little. The work done lifting the atmosphere must come from somewhere. It comes from the thermal energy inside the bag. The air particles inside the bag must lose kinetic energy, becoming colder.

The Learning Activity: Four PEOE cycles are needed for each phase of the investigation:
1. The compression phase.
2. Remove the cork to observe the expansion phase.
3. Add some smoke and repeat the compression phase.
4. Remove the stopper to observe the expansion phase.

Equipment, Preparation and Resources
One plastic carboy with a close fitting stopper. A plastic tube passes through the stopper to the bottom of the carboy. A thermometer is inserted in the stopper (consider using an electronic thermometer).

As you fill the carboy ⅓ full, the air temperature becomes about 7°C warmer. When you remove the stopper, the air cools instantly. No cloud will form, however, unless you introduce some smoke before filling the carboy. Crunch up a bit of paper towel, set it smouldering, and drop it into the carboy before filling with water.

Caution: do not overfill with water. Handle the stopper firmly, so that it does not fly out uncontrolled.

Categories:
Knowledge / Understanding:
Thinking / Inquiry:
Communication:
Applications / Connections:

Assessment and Evaluation
Depth / coherency / accuracy of explanations

Quality of the diagrams and explanations
Any extensions proposed by students, *e.g.*, condensation on dust

10 Academic Science Teachers' Guide
Explaining the Weather

Ms. Boyle: "How do clouds form above the earth?"

"Sure, miss, I can tell you how. Hot air rises! When hot air meets cold air, clouds form."

Ms. Boyle: "Why doesn't the hot air keep on rising, so the clouds form up near space?"

"Well, it must bump into something. Maybe the stratosphere."

Activity 1.10: How Far Will an Air Parcel Rise?

Look once again at the temperature profile of the atmosphere in Activity 1.6. If "hot air is less dense, therefore hot air rises", then why doesn't the relatively warm air at 10 km rise up to 30 km?

Pedagogical Issues Once again, we are dealing with a situation in which everyday experience (smoke rises, clouds form above chimneys) appears to contradict scientific understanding. Cultivating a scientific outlook in our students requires that we get them to value the deeper, more coherent, explanations of scientists. This is certainly an acquired taste!

Science Issues Consider one parcel of dry air, ascending and cooling without exchanging energy with the surrounding air parcels. The rate at which ascending dry air cools relative to the altitude is called the *adiabatic lapse rate*. In the graphs at left, the adiabatic lapse rate is indicated with a dashed line. The local temperature profile is indicated with a solid line. Suppose we heat our dry parcel of air to 30 °C and release it. Consider these scenarios:

A Unstable. Our parcel of warm air ascends and cools, following the dry adiabatic lapse rate. At each altitude, it remains warmer (and less dense) than the surrounding air parcels. It continues to ascend right out of the picture.

When unstable conditions persist, the atmosphere will be in constant motion. Energy will be re-distributed throughout the atmosphere, gradually changing the local temperature profile.

B Stable. Our parcel ascends as before. At an altitude of 2 km however, any further rising causes our parcel to become cooler than the surroundings, and more dense. The parcel then rests at 2 km. This is called *equilibrium*.

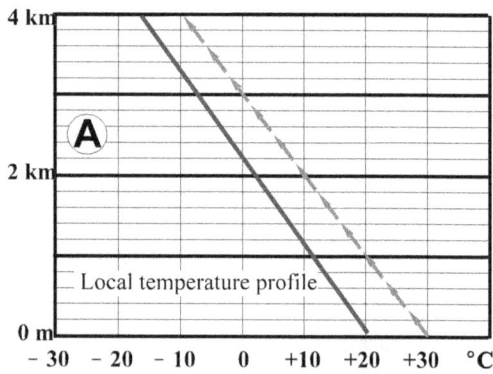

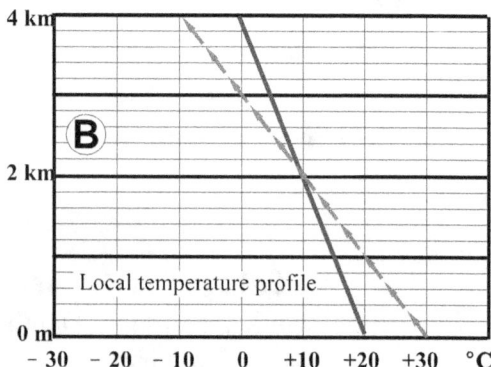

© Ross Lattner Publishing   www.rosslattner.ca

Energy and the Weather

Science and Pedagogy

C   Temperature inversion. For the first 1 km, the local temperature actually gets *warmer* with altitude. The rising air parcel reaches equilibrium at a very low altitude. Temperature inversions are perfectly normal but invisible events. If, however, our parcel contained pollution, then a blanket of pollution would begin to build up over a city. Note that at 12 km, the beginning of the stratosphere, there is a natural temperature inversion. Normally, the air of the troposphere doesn't ascend much beyond this altitude.

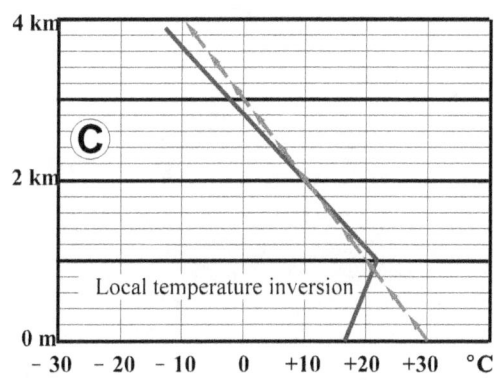

The actual dry adiabatic lapse rate of air near the equator is 9.8 °C / km.

In these lessons, we use an approximation 10 °C/km for ease of calculation. This approximation introduces an error of 2%.

Is an error of this magnitude acceptable when we are learning about weather?

Categories:
Knowledge and Understanding:
Thinking and Inquiry:
Communication:
Applications / Connections:

The Learning Activity

This is a simple pencil and paper activity.

Students calculate the temperature of a dry parcel of air, initially at 30 °C, as it ascends through the atmosphere.

The surrounding parcels of air are at known temperatures. The student's task, then, is to determine if, and at what altitude, the temperature of the ascending parcel becomes less than that of the surroundings, suppressing further ascension.

Equipment, Preparation and Resources

To complete the activity you will need the student exercises, pens, pencils, *etc*.

While the concept is perhaps difficult, the exercise itself is not. You may wish to assign this as a self-study.

Assessment and Evaluation

Correctly calculates the changing T of an ascending parcel

Explaining the Weather

Quiz 1.11: Energy and the Weather

Learning Expectations
ESV.01; ES1.01-05 (stated in prior activities)

Pedagogical Issues
These quiz items are designed to be used in a number of different ways. You could use them as daily quizzes to assess ongoing student understanding. If you use them in this way, score them as formative assessment.

Alternatively, these can be used, or adapted for use, as summative quizzes for general knowledge and application.

Science Issues
The main concepts considered in these quizzes are:

- the relationships among the three ways of transferring heat

- the energy transfers related to changes of state of water

- the relationship between thermal energy and changes in temperature of a parcel of air

- the relationship between the state of a parcel of air (temperature and humidity) and the density of that parcel of air.

These relationships are coherent, and individually quite direct. Together they form a powerful web of ideas that students can employ to explain a wide variety of events.

Energy and the Weather

Science and Pedagogy

The Learning Activity

These quizzes are simple to administer, and should be assessed as quickly as possible after the students attempt them.

Equipment, Preparation and Resources

Students will use the quiz items in the student exercise.

Categories:	Assessment and Evaluation
Knowledge and Understanding:	Correctly answers quiz items
Thinking and Inquiry:	
Communication:	Clarity of responses and diagrams
Applications / Connections:	

10 Academic Science Teachers' Guide
Explaining the Weather

Your students probably already know about simple circulation of the air in convective cells. Unfortunately, the simple convection cell is inadequate to explain the weather.

The participation of water in the convective cell, and the transfer of energy by the change of state of water, plays a determining role in the nature of earth's weather patterns.

A thorough understanding of this water-driven convection cell requires careful application of the particle theory at every stage.

Lab 2.1: The Concept of an Air Cell

Learning Expectations ES 1.04, 1.05: Describe and explain the effects of heat transfer within the hydrosphere and atmosphere on the development, severity and movement of weather systems; explain different types of transformations of water vapour in the atmosphere and their effects.

Pedagogical Issues

All of the concepts that we learned in the last section are applied in this section to describe the circulation of large cells of air. Very few really novel concepts are introduced in this section.

To the professional teacher, then, all of section 2 is an application of previously learned concepts. With that in mind, each of the exercises in Section 2 provides students with an opportunity to display their knowledge and understanding of the concepts learned in Section 1.

To the teacher, each of these exercises affords an opportunity to assess knowledge and understanding, as the students consolidate and integrate their knowledge. It also provides an opportunity to assess the students' application of previously learned knowledge.

Some students will be able to move ahead quite quickly on these exercises with very little guidance. Others will need to be guided to use the concepts which may not have been adequately learned in the first exposure in Section 1.

Science Issues

The convection box experiment is not so much a demonstration of simple convection, as an opportunity for students to depict a circulation cell in terms of the particle theory.

Some of the most important ideas considered are the following:

The energy source has two effects: first, it speeds up the dry air particles, so that they move farther apart. Second, it separates water molecules into the vapour state, with a considerable investment in energy. These changes make the air less dense.

The salt / ice mixture has two effects: first, it slows down the air molecules, so that they move closer together. Second, the slower water molecules stick together, giving off the energy they were given by the heater. These changes make the air more dense.

Air Circulation

Science and Pedagogy

There are only four forces in the universe.*

The electric force is providing the energy to move the air and water particles apart at the hot plate.

The electric force is also responsible for moving the particles back together at the ice cans.

The gravitational force can only pull matter toward the earth. It is worth emphasizing that gravity exerts the strongest pull on the greatest concentrations of matter.

In this case, the greatest concentrations of matter are found in the denser, (i.e. colder, drier) air.

*Other forces have been suggested, e.g. "dark matter."

Stay tuned.

The Learning Activity
Before the experiment, students will
- **Predict**: the direction of the air flow.
- **Explain**: their prediction, using particle diagrams and sentences.

Students will conduct the experiment and
- **Observe**: how the air circulates.
- **Explain**: the circulation, using the particle theory, energy of change of state, modes of heat transfer, and other ideas you learned in this unit

Equipment, Preparation and Resources
Each group of 2-4 students needs:

- one cardboard box
- two or three pop cans
- hot plate
- beaker
- test tube
- incense
- ice and salt

To avoid possible hazards ensure that:

- hot plate should be kept low
- ice/salt mixture can get very cold, freezing of skin is possible

Categories: Assessment and Evaluation
Knowledge and Understanding:	Quality, depth, breadth of knowledge of particles, forces, etc.
Thinking and Inquiry:	Successfully completes the PEOE cycle
Communication:	Clarity of writing
Applications / Connections:	Applications of the particle theory, energy transfer, *etc*.

Explaining the Weather

10 Academic Science Teachers' Guide

As scientists, we use representations of nature to solve problems of interest to us and to our community.

As science teachers, we initiate our students into the construction and use of specialized representations of nature.

When we are considering teaching a particular representation to our students, we need to ask two questions:

What problems does this representation equip our students to solve?

Could the community of scientists use this representation to solve a problem today?

If a particular representation does not help real scientists and real students to solve real problems, then why should we teach it?

Activity 2.2: Air Cells and the Formation of Cumulus Clouds

Learning Expectations ES 1.04, 1.05: Describe and explain the effects of heat transfer within the hydrosphere and atmosphere on the development, severity and movement of weather systems; explain different types of transformations of water vapour in the atmosphere and their effects.

Pedagogical Issues

Lab 2.1 was an analogy to the situation described in Activity 2.2. This exercise is set up as a series of discrete steps that guides students to apply their "school knowledge" to a very commonly observed, but little understood phenomenon. Students will likely find this very challenging at first. We might expect several dead ends.

A student who has inadequately learned the concepts in Section 1 may not be in a position to draw upon that knowledge to complete the required analysis.

A student who has very vivid "everyday notions" of clouds may draw upon those ideas instead of the scientific explanation.

A student who is not accustomed to making frequent, varying judgements may find the task quite unpleasant or incomprehensible.

Science Issues

If one of the objectives of science is to achieve a coherent explanation of the universe, then we ought to be able to explain cloud formation under the same terms as we explain everything else. It is essential, then, that we not short circuit the process by substituting less coherent explanations when a really strong explanatory framework is available to the students.

Air Circulation

Science and Pedagogy

The Learning Activity
The following are answers to the questions in the student exercises:

1. a. At the original conditions, the air parcel was 10 °C. At that temperature, it could contain 7.6 g of water, but only contained 20% of that amount, or 1.52 g water. At 26 °C, its water content increased to 10.65 g. The mass of water added was 9.1 g.
   b. It takes 2500 J to vaporize 1.0 g of water. The energy to vaporize 9.1 g of water is 23 000 J. It also takes 1000 J to heat 1.0 kg of dry air through 1 °C. Raising the temperature of the dry air took 16000 J. Total energy input: 39 000 J.
   c. Compared to 0 °C, the 10 °C air parcel would be 27.5 mm taller. The additional 1.5 g of water would make the parcel another 0.75 mm taller. Total height of the air parcel is 773 + 27.5 + 0.75 = 801 mm. The density of the air parcel then is 1.00 kg / 0.801 m^3 or 1.25 kg/m^3 At the higher T and RH, density is 1.00 kg / 0.850 m^3 = 1.18 kg/m^3.

2. a. The air parcel in 1 contained 10.65 g of water. The temperature at which 10.65 g of water is the saturation point is 15 °C. Water will begin to condense when the air cools to 15 °C.
   b. Since the air starts at 26 °C, and cools by 1.0 °C for ever 100 m ascended, the air will cool to the dew point at 1100 m. So the base of the cumulus cloud will be 1100 m.

3. a. At –2 °C, only 3.3 g of water is left in the air parcel. All the rest of the 10.65 g of water must have condensed out. That is 7.3 g of water.
   b. The condensation of 7.3 g of water would release 18300 J of heat.

4. a. 18300 J of heat could heat 1.00 kg of dry air by 18.3 °C. The temperature of the air parcel would actually be 16.3 °C. This is very unlikely.

5. a. The air parcel has actually been radiating away most of the thermal energy given to it by the condensing water.. When it cools to –2 °C and 50% RH, it will contain 1.6 g of water.
   b. Its density would be 1.00 kg / 0.768 m^3, or 1.30 kg/m^3. This is denser than the air on the ground, and so our air parcel begins to descend.

6. a. As our air parcel, containing 1.6 g of water, descends from 2.0 km to ground, it heats up due to compression. The total heating is 2.0 km × 10 °C , or 20 °C. The temperature of our air parcel on the ground is then 18.0 °C.
   b. At that temperature, 1.6 g of water represents only 12% of the maximum at 18 °C. So the %RH of the freshly descended air parcel is only 12%. Ready for another trip!

This overall set of calculations is quite involved, yet each step is very simple and direct. We have avoided any mention of simultaneous events such as the radiation of heat energy during the process of rising and condensation of water. This is beyond the mathematical ability of most of us, and yet the overall conclusions of our stepwise set of calculations is a close approximation of the cumulus air cell.

Categories: Assessment and Evaluation
Knowledge and Understanding: Correct execution of all calculations
Thinking and Inquiry:
Communication: Clarity of expression of the chain of reasoning
Applications / Connections: Appropriate and useful applications of concepts and math

Explaining the Weather

Do you remember your first high school geography class? Can you recall the treatment of the idea of a "rain shadow"?

Definition: "Rain Shadow: a very low rainfall on the leeward side of a mountain."

Is it likely that a student, or a scientist, could solve any real problems related to weather on the basis of a definition of that sort?

Are definitions the best we can do to initiate our students into the practice of science?

Activity 2.3: The Chinook

Learning Expectations ES 1.04, 1.05: Describe and explain the effects of heat transfer within the hydrosphere and atmosphere on the development, severity and movement of weather systems; explain different types of transformations of water vapour in the atmosphere and their effects.

Pedagogical Issues

If a student had difficulty with the last exercise, then this activity provides an opportunity for students to improve their skills.

The basic processes in the chinook are similar to the formation of cumulus clouds, but the overall phenomenon is quite different. The representational, conceptual and procedural knowledge that the students gained in Section 1 - Energy and the Weather - can be applied to this novel situation. Those students who were successful on the last exercise will learn about a new weather phenomenon. Those who were not successful can correct some of their approaches and strengthen their skills.

Science Issues

The chinook is a complex phenomenon, with many relationships to high school studies in geography, history, agriculture, politics, anthropology and more.

The rain forests of the west coast, the rain shadow in BC's interior, and the chinook wind itself have all shaped the modern history, and pre-history, of the west.

This is also the first place in this course in which we have made a connection between the local geographic features and weather patterns. The reason we have delayed is simple, but it is important. Local weather phenomena involve complex interactions among movements of air, of water, and of energy. If students (or scientists) lack intimate knowledge of the nature of those processes, then the students lack the ability to construct an explanation. In such a world, all that we can do as teachers is "tell them" about the rain shadow. Surely this is not adequate?

Air Circulation

Science and Pedagogy

The Learning Activity

The following are answers to the questions in the student exercises:

1. a. An air parcel over the Pacific at that time of year is 16 °C and 100 %RH. It would hold 11.3 g of water vapour.
   b. A dry air parcel at 0 °C would have a height of 773 mm.
      Warming the parcel to 16 °C would increase its height by 44 mm (16 × 2.75 mm).
      Adding 11.3 g of water would increase the height by an additional 5.7 mm (11.6 × 0.50)
      The total height of the parcel would be 773 + 44 + 5.7 = 823 mm = 0.823 m.
      The volume of the air parcel would be 0.823 m^3
      The relative density of the air parcel would be 1.20 kg/m^3

2. a. At 3.0 km, the parcel would have cooled by 30 °C. Since it started at +16 °C, its temperature at that altitude would be –14 °C.

3. a. When 6.0 g of water condenses, it releases (6.0 ×2500) 15 000 J of energy.
   b. Recall that the specific heat capacity of dry air is 1000 J/g•°C. That is, 1000 J will heat a parcel of air by exactly 1 °C. If all 15 000 J of energy are used to heat the air, the temperature change will be 15 °C. Since we started at –14 °C, the temperature in the rain clouds would be +1 °C, just above freezing.

4. a. When water vapour changes to ice, it releases 2830 J per gram. When 5.0 g of water freeze out as snow, the water loses a total of 14 150 J.
   b. If all 14 150 J of heat energy released by the formation of snow are used to heat the air, the temperature change of the dry air will be 14.15 °C. The air temperature was initially –33 °C, so the final temperature of the parcel at the top of the mountain is –19 °C

5. a. As the parcel falls from 5 km to 1 km, its temperature will increase by 40 °C. Since it started at –19 °C, its temperature will be 21 °C

6. a. An air parcel at 21 °C is capable of holding 15.7 g of water. Our parcel only contains 0.3 g of water, or 2 % of the maximum. The air is very dry at 2 %RH

Equipment, Preparation and Resources

Pens, paper, calculators, and the exercises from the student manual.

Categories:	Assessment and Evaluation
Knowledge and Understanding:	Correct execution of all calculations
Thinking and Inquiry:	
Communication:	Clarity of expression of the chain of reasoning
Applications / Connections:	appropriate and useful application of concepts and math

10 Academic Science Teachers' Guide

Explaining the Weather

Ms. Boyle: "What makes a thunderstorm?"

Allen: "I know, I know... We get thunder and lightning and rain when a hot air mass and a cold air mass collide."

Ms. Boyle: "How do the colliding air masses make thunder and lightning?"

Allen: "We can hear them... kind of... crashing together.... and that makes lightning."

Brenda: "No, rainy weather is caused by low pressure zones moving in from somewhere else."

Ms. Boyle: "Where did you learn that?"

Brenda: "That's what they say on the weather channel."

Activity 2.4: Air Masses and Atmospheric Pressure

Learning Expectations ES1.04, 1.05: Describe and explain the effects of heat transfer within the hydrosphere and atmosphere on the development, severity and movement of weather systems; explain different types of transformations of water vapour in the atmosphere and their effects.

Pedagogical Issues

The purpose of the unit so far has been to provide the student with the means to represent changes in the atmosphere at two distinct levels: the level of particles, and the level of the air parcel. At this point, we move to a larger unit of analysis, the air mass.

Students often think of air masses as distinct things, which have motivations and movements all their own. We must take care to describe air masses in terms of the explanatory framework we have already put in place, that is, particles and parcels of air.

Another prominent area for student misunderstanding is the notion of air pressure. Students think of pressure as the cause of weather events. For example, a student is likely to say that "high pressure causes clear skies" or "low pressure will cause rainy weather." Children (and adults) tend to use linear cause-and-effect reasoning as a first resort. When confronted with two simultaneous phenomena, such as low pressure and clouds, human beings are very likely to say that one causes the other.

The problem is that they are partly, but only partly, right. On the one hand, the conditions which give rise to low pressure, i.e. warm moist air masses overhead, are precisely the conditions that prevail during rains. On the other hand, a low pressure system moving overhead *does* cause changes in local air parcels.

Science Issues

As a mass of warm, moist low density air passes overhead, all of the parcels near ground respond to the lower pressure by expanding. But expansion causes cooling. Cooling increases relative humidity. Cooling also increases the likelihood of the formation of clouds near the ground (fog). Conversely, when a mass of cold, dry high density air passes overhead, all of the parcels near the ground respond to the higher pressure by collapsing a little. The compression causes warming, lowers the relative humidity, and induces fog to evaporate, thus clearing the air. These changes can occur in seconds, without the presence of a wind!

Air Circulation

Science and Pedagogy

The Learning Activity

This activity is also a guided calculation activity, to explore the relative effects of air masses passing overhead.

For example, if we are sitting under a cool dry air mass, and the pressure outside is 1025 mb, there must be a pile of 10 250 air parcels sitting overhead (assuming g = 10 m/s² See note at left).

Likewise, if we are sitting under a warm moist air mass, and the pressure outside is 990 mb, there must be a pile of 9900 air parcels sitting overhead.

Note that it is sunlight which provides the kinetic energy that inflates the warmer parcels, so that each kilogram of air occupies a larger volume at any given altitude and pressure.

Over a period of hours, the cool, dry, dense air will settle to the bottom, spreading out. The warm, moist, less dense air will be forced up and over, filling in the region vacated by the descending cool air.

For ease of calculation, we are using the approximation that the acceleration due to gravity, g = 10 m/s². The accepted value is actually g = 9.81 m/s².

We have therefore knowingly introduced an error of 2%.

Given all of the other variables unaccounted for in our study of the weather so far, is our approximation likely to cause any fundamental errors of understanding?

In general, if the pressure upon an air parcel increases 10 mb, compression causes the temperature of the parcel to increase by 1 °C.

Conversely, a decrease in ambient pressure of 10 mb allows an air parcel to expand, and to become 1 °C cooler.

Equipment, Preparation and Resources

Pens, paper, calculators and the student exercises from the lab manual are necessary for this activity.

Assessment and Evaluation

Calculates changes in temperature due to changes in pressure

Categories:
Knowledge and Understanding:
Thinking and Inquiry:
Communication:
Applications / Connections:

Explaining the Weather

10 Academic Science Teachers' Guide

> Warm fronts come through your area like a butter knife gliding over butter: the knife (warm air) rides up and over the butter, pushing the butter (cold air) along beneath it.

> Application of knowledge is a complex mental phenomenon. It does not just magically appear in children.

> Students must repeatedly tackle challenging problems with a powerful set of conceptual and procedural tools.

Activity 2.5: Warm Fronts

Learning Expectations ES1.04, 1.05: Describe and explain the effects of heat transfer within the hydrosphere and atmosphere on the development, severity and movement of weather systems; explain different types of transformations fo water vapour in the atmosphere and their effects.

Science Issues

We often represent warm fronts in our books and learning materials as having fairly steep angles, 20° or more. In fact, the slope of a warm front is very shallow, only a degree or two. The high front edge of a warm front may be right over our heads, while the ground-level edge of the warm front is still 300 km away.

The first signs of an approaching warm front are *cirrus* clouds, wispy mare's tails composed of ice crystals much finer than terrestrial snow, some 8 km overhead. This may occur one or two days before the rest of the warm front comes through, bringing cloud, drizzle and rain.

Our work with the chinook should prepare students to undertake the appropriate analysis here. Note that the warm air mass is being forced up and over the cold air mass, much like the Pacific air was forced up and over the Rocky mountains.

The clouds that form during a warm front are not at all like cumulus clouds. Because the warm, moist air is moving up and over a gentle slope, the clouds form in layers, or *strata*. The *stratus* clouds are likely to be low, 1000 m or less, and to shed rain (*nimbostratus*).

Pedagogical Issues

Once again, we are using the conceptual and procedural knowledge covered in Section 1 - Energy and the Weather - to analyze the complex events inside a warm front.

The warm front that we describe is composed of air parcels, whose particles behave in ways that should now be familiar to students.

Air Circulation

Science and Pedagogy

The Learning Activity

The following are answers to the questions in the student exercises:

1. a. Two different air parcels in the warm and cold air masses will have quite different characteristics.

    The warm air parcel, 20 °C and 60 % RH will contain 8.8 g of water.
    The height of the warm air parcel is 773 + 55 + 4 = 832 mm.
    The volume of the air parcel is 0.832 m³.
    The density of the 1.00 kg air parcel is 1.20 kg/m³.

    The cool air parcel, 6 °C, and 20 %RH will contain 1.16 g of water.
    The volume of the air parcel is 0.790 m³.
    The density of the 1.00 kg air parcel is 1.27 kg/m³.

    The cooler air parcel is more dense, and the warm moist air in the warm front is pushed up and over the cool air mass as the warm front advances.

2. a. Each warm air parcel contains 8.8 g of water. An air parcel would be saturated by 8.8 g of water at the temperature of 12 °C. A warm air parcel would have to cool from 20 °C to 12 °C in order to reach the dew point, a temperature change of 8 °C. The warm air parcel could reach that temperature if it ascended to 800 m, and cooled by expansion. We would expect to see clouds begin to form at 800 m.

3. a. As our warm air parcel is pushed farther to the right and higher, it cools by 10 °C per km. At an altitude of 2.0 km, its temperature would have fallen by 20 °C by expansion. We would expect the temperature to be 0 °C. At that temperature, a parcel of air could contain only 3.8 g of water. Since the parcel started with 8.8 g of water, then 5.0 g of water must have precipitated out as rain.

4. a. A cool parcel is at 6 °C and could hold a maximum of 5.8 g of water. Since it already holds 1.2 g of water, it could easily absorb 1.20 g more.
    b. The heat needed to vaporize 1.20 g of water is (2500 × 1.20) = 3 000 J.
    3. The air parcel would cool by 3.0 °C in order to vaporize 1.20 g of the falling rain. The final temperature of the air parcel would then be 3.0 °C. The final humidity would be 50%

5. a. The warm front causes the pressure to fall from 1005 mb to 995 mb, a difference of 10 mb. The air parcels close to the ground will expand, causing them to cool by yet another 1.0 °C. Temperature of the air parcel is now 2.0 °C, and the air becomes even foggier.

6. a. A person on the ground would see *cirrus* clouds (mare's tails) high overhead as the first signs of an approaching warm front.

Categories: Assessment and Evaluation

Knowledge and Understanding: Correct use of conceptual and procedural knowledge
Thinking and Inquiry: Good judgement in choosing approaches to the problems
Communication: Quality of writing, diagrams, calculations, *etc.*
Applications / Connections: The exercise as a whole is an application of Section 1

Explaining the Weather

10 Academic Science Teachers' Guide

Cold fronts come through your area like a bulldozer blade:

the blade (cold air) cuts under the soil, lifting the soil (warm air) and pushing it up and ahead.

Activity 2.6: Cold Fronts

Learning Expectations ES 1.04, 1.05: Describe and explain the effects of heat transfer within the hydrosphere and atmosphere on the development, severity and movement of weather systems; explain different types of transformations fo water vapour in the atmosphere and their effects.

Pedagogical Issues
The difference between a warm front and a cold front may appear obvious:
- warm front - warm air is pushing upon the cool air,
- cold front - cold air is pushing upon the warm air.

However, the difference in representation is subtle.

Science Issues
In the case explored here, the moving cold front is pushing a very warm and moist air mass to a high altitude. This process results in the development of a thunderstorm. In the very early or very late winter, when significant amounts of water vapour are present in the atmosphere, these conditions can result in blizzards.

The condensation of water vapour provides thermal energy to the growing cumulonimbus cloud (thunderhead), so that the rising air is continually warmer than its surroundings, up to very high altitudes.

This process is common to a number of weather phenomena. Other aspects of thundershowers can be left until the students undertake individual projects on extreme weather phenomena.

The Learning Activity
The following are answers to the questions in the student exercises:

1. a. The warm air parcels contain (70% × 27.2 g) 19.0 g of water; the air parcels in the cold front contain (20% × 8.7 g) 1.74 g of water.

   b. The height of the warm air parcel is. 773 + 82.5 + 9.5 = 865 mm.
   The volume of the air parcel is 0.865 m^3.
   The density of the 1.00 kg air parcel is 1.16 kg/m^3.
   The volume of the cold air parcel is (773 + 33 + 1 = 806) 0.806 m^3.
   The density of the 1.00 kg air parcel is 1.24 kg/m^3.

   c. The moving air parcel in the cold front is more dense than the warm moist air. The colder, drier, denser air parcel sinks under the warm parcels, and pushes the warm moist air parcels up and over the colder air mass as the cold front advances.

© Ross Lattner Publishing — www.rosslattner.ca

Air Circulation

Science and Pedagogy

2. a. If the warm air parcel was lifted to 3.0 km, it would cool by 30 °C, to T = 0 °C.
   b. At 0 °C, the air parcel could only hold 3.8 g of water. Therefore, as much as 15.2 g of water could condense out as cloud droplets.
   c. If 8.0 g of water condensed out as cloud droplets in each air parcel, 20 000 J of heat energy would be released by condensation. This energy would heat the 1.00 kg air parcel by 20 °C
   d. The air parcel would rise in T from 0 °C to 20 °C.

3. a. As the parcel ascended from 3.0 km to 8 km (i.e. 5.0 km), expansion would cause it to chill by 50 °C to –30 °C.
   b. The remaining 11 g of water would freeze. Each gram of freezing water vapour would release 2830 J of energy, or 15 300 J of thermal energy altogether.

4. a. The air parcel, at –30 °C is still warmer than its surroundings. It would continue to rise to the edge of the stratosphere, carrying with it fine ice crystals. At that altitude, it encounters the strong winds near the stratosphere which blow the ice crystals into the smooth, flat, slightly rounded profile of the anvil cloud. Above this altitude, the temperature of the surroundings increases with greater height (see diagram). This is the natural temperature inversion that marks the edge of the stratosphere. Our air parcel will not rise beyond this altitude. If it did ascend further to 16 km, its temperature would cool to –100 °C or less, due to expansion. At that temperature, it would then be colder than its –30 °C surroundings, and would sink.

5. a. The size of a thunderstorm is absolutely enormous. If we assume a thunderstorm 2 km wide, but 10 km long, by 10 km high, we are talking about a total volume of 2×10^{11} m^3. The mass of air parcels lies in the range of 1×10^{11} kg. If each of the original warm, moist air parcels contained 27.2 g of water vapour, and all of it was condensed as either ice or water, then the total mass of water available would be $(1 \times 10^{11})(0.0272$ kg$) = 2.7 \times 10^9$ kg, or 2.7 million tonnes of water. Perhaps half of this actually falls as rain.

6. a. The muggy air is gone. A fresh blast of cold air from up near the stratosphere was carried down to you by a million of tonnes of falling rain. You saw gusts of cold, clean air blow the leaves of the trees around you, and felt the air on your skin. The thunderstorm has moved on, and the cold front has passed overhead. Now as the mass of air over your head increases, the pressure on each parcel near the ground begins to increase rapidly as well. This increase in pressure causes the air parcels near the ground to become warmer, drier, and clearer. Better days are ahead!

Categories: **Assessment and Evaluation**
Knowledge and Understanding: Correct use of conceptual and procedural knowledge
Thinking and Inquiry: Good judgement in choosing approaches to the problems
Communication: Quality of writing, diagrams, calculations, *etc.*
Applications / Connections: The exercise as a whole is an application of Section 1

10 Academic Science Teachers' Guide
Explaining the Weather

How many concepts must be kept "alive" in one's head when one is attempting to learn the notion of an isobar?

At least 6... This must approach or surpass the limit of human cognition. The only way to keep them in front of the student's mind is to provide a representation that keeps it all in front of the student's senses.

Activity 2.7: Atmospheric Pressure and Isobars

Learning Expectations ES1.06: Describe the factors contributing to earth temperature gradients and to wind speed and direction.

Pedagogical Issues

We continue to apply the concepts from section 1 - Energy and Weather - to new situations. The "occluded front" is a new concept, built upon the previous concepts of "warm front" and "cold front". We are further introducing the concept of an *isobar*, a line of equal atmospheric pressure on a map.

This is a very complex set of concepts. It is accompanied by a complex set of representations.

The "warm front" and "cold front" representations of the previous exercises are combined. The "stack of super-parcels" is introduced. Finally, this combined representation is related to a flat weather map.

Science Issues

An attempt has been made here to use the concept of an air parcel to provide reasonable numbers to support student learning of the concept of an isobar. A parcel of air is too small to use effectively here, so we are combining 500 parcels into a "super-parcel". The countable number of super-parcels in each stack of air make the exercise relatively simple.

On the other hand, the numbers that appear here are quite extreme - 950 mb and 1050 mb are definitely outside the normal range of atmospheric pressures.

The concept of an isobar will become central in the next section, as we attempt to have students predict the movement of air between high pressure and low pressure zones, as both the zones and the earth are spinning.

The Learning Activity As in the previous exercises, the students answer a series of questions about the movement of air parcels within an occluded front. The following are answers to the questions in the student exercises:

Air Circulation

Science and Pedagogy

1. Height of air cool parcel = 773 + 36 = 809 mm
   Volume of cool air parcel = 0.809 m^3
   Density of cool air parcel = 1.00 kg ÷ 0.809 m^3 = 1.24 kg / m^3

2. Height of cold, dry air parcel = 773 + 14 = 787 mm
   Volume of cold, dry air parcel = 0.787 m^3
   Density of cold, dry air parcel = 1.00 kg ÷ 0.787 m^3 = 1.27 kg / m^3

3. Height of warm air parcel = 773 + 71.5 = 845 mm
   Volume of warm air parcel = 0.845 m^3
   Density of warm air parcel = 1.00 kg ÷ 0.845 m^3 = 1.18 kg / m^3

4. The warm moist air is less dense than either of the other air masses. The cold front will sink beneath all of the air of lesser density.

5. Stack X is the most massive, Stack Y is the least massive. Stack X is certainly descending, and stack Y is certainly ascending.

6. The pressure under stack X is 1050 mb, under stack Y is 950 mb and under stack Z is 1000 mb. These figures are extreme pressures, but the representation is "do-able".

Equipment, Preparation and Resources

Prepare photocopies of the occluded front diagram from the student exercises. Each student work group can cut the occluded front diagram and fasten it to the map to provide a three dimensional image as shown below. The isobars represent stacks of air parcels of similar density.

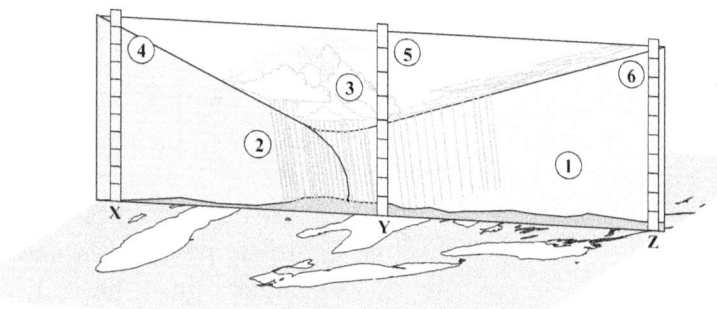

Categories:	Assessment and Evaluation
Knowledge and Understanding:	Correct use of conceptual and procedural knowledge
Thinking and Inquiry:	Good judgement in choosing approaches to the problems
Communication:	Quality of writing, diagrams, calculations, *etc*.
Applications / Connections:	The exercise as a whole is an application of Section 1

Quiz 2.8: Air Circulation
Answers to quiz items

1	Three similarities: a) On the right hand side, the particles are farther apart, they are moving faster, and there are more water particles. b) On the left hand side, the particles are closer together, they are moving more slowly and have fewer water particles. c) The air rises on the right and sinks on the left.	2	In general, this idea is true. Air doesn't just rise. Air parcels are forced upward by other air parcels that are falling somewhere else.
3	Both air parcels are at the same altitude, so have the same pressure. The only difference is their relative temperature. Air parcel A inside the cloud is slightly warmer than air parcel B in the blue sky outside the cloud. Therefore, parcel B will sink, pushing parcel A up.	4	If you extend the situation in (3) to include the number of sinking air parcels in the clear air around a cumulus cloud, then the large number of sinking air parcels must exert an upward force upon a warm, moist air parcel of lower density. Most naive accounts of air circulation suggest that "hot air rises". Look how the picture changes when we focus our attention on cold, dense air being pulled downward by gravity.
5	The problem with the "1 cell" model: hot tropical air lifted to high altitudes and chilled by expansion must remain aloft for a journey of 9000 km to the poles. This is very unlikely.	6	The problem with the "2 cell" model: Suppose the cold air begins to sink about half way to the poles. To keep the directions of flow synchronized, we would have to have frosty polar air ascending above the poles. Again, this is unlikely.
7	What about the "3 cell" model? Tropical air ascends near the equator, travels 3000 km, and descends about 30° latitude. The descending cold air travels another 3000 km, finally ascending about 60° latitude. The last cell descends over the poles. That makes sense!	8	As the earth tilts toward and away from the sun with each season, the exact location of the equatorial Hadley cells shifts north or south on the globe. The cold air still descends over the poles. This means that the Hadley cells actually get alternately longer and shorter in the northern and southern hemispheres.

Air Circulation

Science and Pedagogy

9 The air parcel is dry, containing only 10% of the possible 30.7 g of water, or 3.1 g of water. The air parcel rises to 1 km, cools to 22°C. Water does not condense, as the 3.1 g of water vapour is less than the 16.7 g of water it could hold at that temperature. The parcel is warmer than the surroundings, and continues to rise. The parcel rises to 2 km, cools to 12°C. It is now cooler than the surroundings, and can rise no farther. Still no condensation or cloud.	**10** This air parcel is warm and moist, containing 19.0 g of water vapour. It rises to 800 m, cools to 24°C, and condensation begins. Each parcel will get warmer by (1.0 × 2500 ÷1000) = 2.5 °C for each gram of condensation. The air parcels will continue to rise to about 3 km., as white clouds that radiate heat to the surrounding air.
11 The warm, moist air causes large waves or "hills" of cold air. As the warm, moist air rises up over each "hill", a line of clouds forms on the windward side of the hill. As the air descends on the other side of the "hill", it warms, the cloud evaporates, and a clear line develops. The clouds look like rows of cloudy ripples, all at the same altitude.	**12** There are cirrus clouds, very high ice clouds. This means two things: 1. moisture has been lifted to great altitudes. Water vapour probably permeates the entire atmosphere from the ground up to those clouds. 2. High altitude ice crystals, falling through cold, high-altitude cumulus clouds, can trigger the formation of large drops of water that will fall all the way to the ground. They indicate rain tomorrow or next day!
13 These ripples of clouds are described in (11) above. A warm front is moving in, and we will see stratus clouds tomorrow, with a chance of rain as well.	**14** A wall of clouds is what you see when a cold front approaches you. A cold front from the east is common in winter, and brings with it two things: Cold air, of course, but also air saturated in Atlantic moisture. We are getting snow tonight, maybe even a storm.
15 A cold air mass has moved in, bringing cold, dry air from the northwest. The increasing pressure causes the cold air mass to warm slightly. Sunshine reaches the earth unimpeded, so that the earth feels warm. Yet because the temperatures are cool, convection is suppressed. Enjoy the still, warm autumn air!	**16** A wall of clouds is a sign of an approaching cold front. We are in for some very big thunderstorms. Think of the thermal energy of the water vapour; it will all be used to provide energy to drive the violent air movements of the coming summer storm.

10 Academic Science Teachers' Guide
Explaining the Weather...

Should we call it the "Coriolis force" or the "Coriolis effect?" If we call it the "Coriolis force," we identify it as a real force. Most scientists agree that there are only four forces. The "Coriolis force" is not among them. If we call it the "Coriolis effect," we may diminish its role as a *cause* of observed weather.

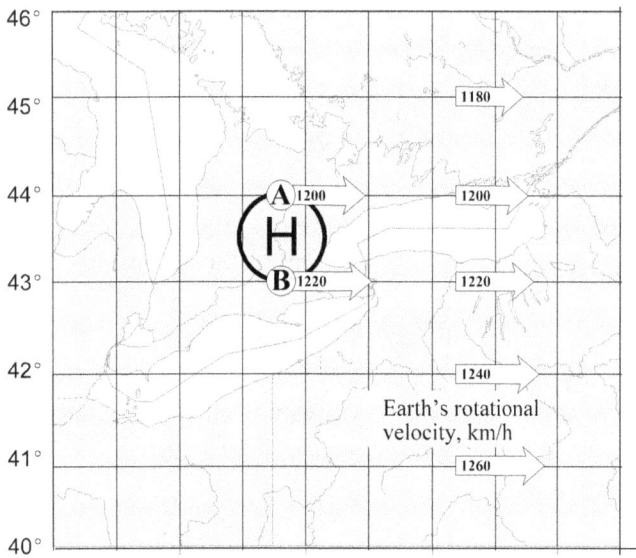

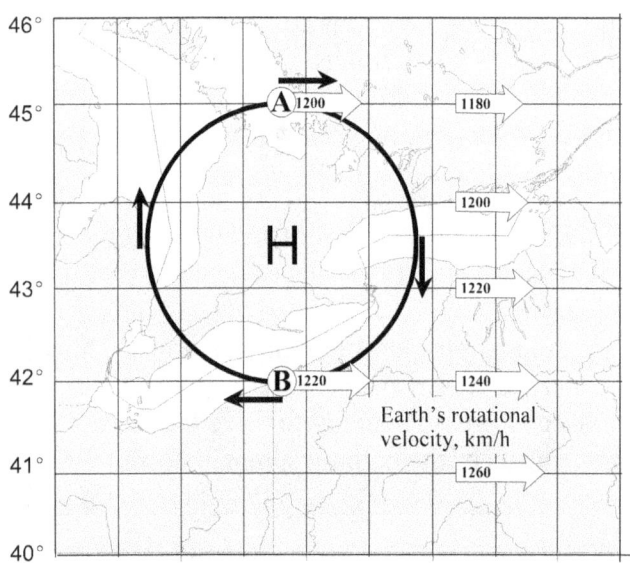

Activity 3.1: We Live on a Spinning Ball

Learning Expectations ES1.06: Describe the factors contributing to earth temperature gradients and to wind speed and direction.

Pedagogical Issues
The main concept here is the Coriolis effect, that is, the apparent motion of air and water currents due to the spinning of the earth. There are two dominant approaches in school texts: some call it the "Coriolis force"; others call it the "Coriolis effect". Both of these approaches appear to create some learning difficulties.

We will first of all avoid naming the phenomenon until students understand the principles behind it. Instead, we will try to describe it in terms of more fundamental ideas. Two different approaches will be used. The conservation of momentum principal underlies both of these approaches.

1. **Conservation of Linear Momentum**
   Examine the diagram above left. Newton's first law might say: "If a 1.00 kg parcel of air A is in motion at 1200 m/s [E], it will continue at that speed and direction unless acted upon by an external unbalanced force." Since there is very little frictional force of any kind to change the speed and direction of our parcel of air, it does indeed keep moving at 1200 m/s [E], even as it drifts slowly north.

   Twenty hours later, we are in the situation in the diagram below left. Parcel A is still moving to the east at 1200 km/h, but the earth at that latitude is only moving 1180 km/h. Parcel A is moving 20 km/h [E] *faster* than the earth. We would experience it as a wind, moving 20 km/h [E]. Likewise, Parcel B is moving 20 km/h slower than the earth. We experience it as a wind moving 20 km/h [W]. The whole air mass is spinning clockwise, relative to the earth.

... on a Spinning Planet

Science and Pedagogy

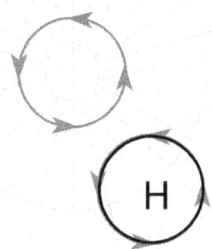

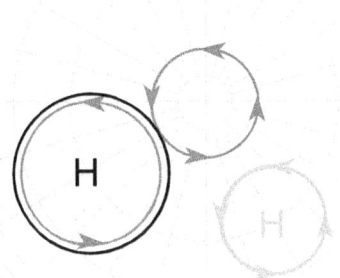

2. Conservation of Angular Momentum
In the diagram above left, we are looking down upon the North pole, and we see the earth rotating toward the east. The high pressure zone H is rotating at exactly the same angular velocity as the earth.

Twenty hours later, the high pressure zone has moved with the earth to the new situation. The high pressure zone has expanded, but its angular momentum is conserved. Because the high pressure zone is larger, its moment of inertia is larger, so its angular velocity is decreased. The high pressure zone is spinning at a lower angular velocity than the earth. To a person standing on the earth, the air mass is spinning clockwise.

Try this demonstration. Put a student on a rotating chair. Give the student some heavy masses to hold. Get the student spinning, and have the student alternately extend and retract their arms. Watch what happens.

The Learning Activity
Students work through four prediction exercises, the first with guidance and the last three without.

1. Low pressure zones spin counterclockwise (same as the earth seen from the North Pole).

2. The system is a cyclone, developing into a hurricane. The low pressure zone in this example would be spinning at 87 km/h, if there was no friction to slow it down!

3. The system over Newfoundland will develop winds blowing to the southeast.

Equipment, Preparation and Resources 1 rotating stool, and two heavy barbells or other easy-to-grip masses.

Categories:
Knowledge and Understanding:
Thinking and Inquiry:
Communication:
Applications / Connections:

Assessment and Evaluation
Correctly predicts the rotation of pressure systems

Explaining the Weather

10 Academic Science Teachers' Guide

In weather lore, an "east wind" is a wind which blows *from* the east. In no other area of science do we name a velocity according to where it came *from*. We will speak of an eastward wind, or a westward wind, indicating where the wind is *going*. In this way we will remain coherent with the rest of the scientific endeavor.

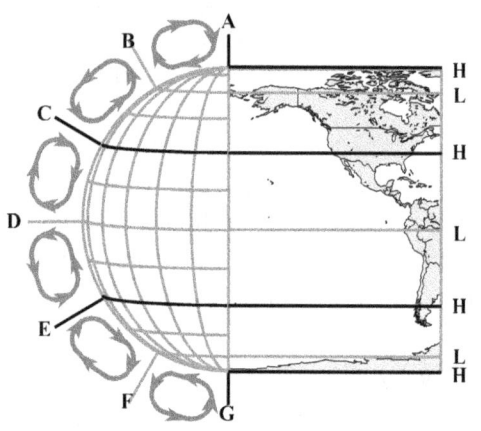

Activity 3.2: Hadley Cells and Prevailing Winds

Learning Expectations ES1.06: Describe the factors that contribute to earth temperature gradients, and to wind speed and direction.

Pedagogical Issues

In this activity, we are supporting student attempts to relate the Hadley Cells to easterly and westerly prevailing winds. We start by reviewing the idea of Hadley cells, and relating it to high and low pressure systems. The main concept is that atmospheric pressure is low under ascending air, and high under descending air columns.

Science Issues

The Hadley cell system creates alternating bands of high and low pressure at specific latitudes. Generally, there is a permanent low pressure zone near the equator and at 60° in the north and south hemispheres. There is descending air, and a permanent high pressure zone, at the poles and at 30° in both hemispheres.

Hadley cells create northward and southward air currents at the earth's surface. If the earth did not spin on its axis, these winds would blow due north and south.

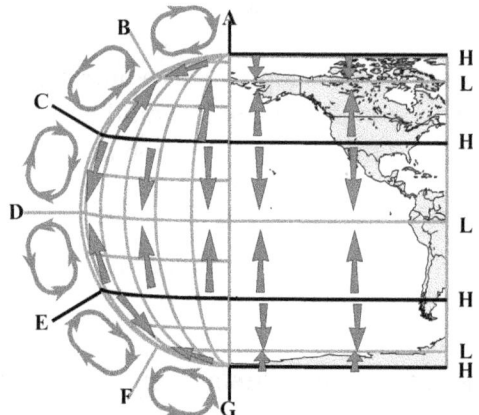

© Ross Lattner Publishing   www.rosslattner.ca

... on a Spinning Planet

Science and Pedagogy

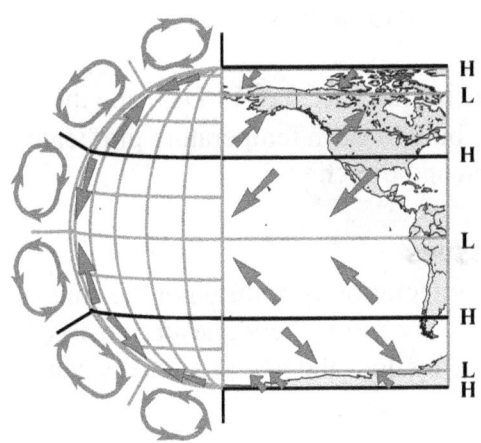

Because the earth spins, poleward moving winds also maintain their eastward velocity, and tend to become strong eastward winds.

Air that drifts toward the equator originates at more slowly moving latitudes on the earth's surface. These tend to become relatively slower than the earth's surface as they drift toward the equator. These currents become westward moving winds.

The Learning Activity

The activity itself is quite brief. Students are to draw arrows on the maps to indicate the movement of air due to the Hadley cell circulation.

Equipment, Preparation and Resources
- copies of the student exercise
- pens, pencils, *etc*.

Categories:
Knowledge and Understanding:
Thinking and Inquiry:
Communication:
Applications / Connections:

Assessment and Evaluation
Correctly draws arrows indicating prevailing winds

10 Academic Science Teachers' Guide
Explaining the Weather

Activity 3.3: The World in January
Activity 3.4: The World in July

Learning Expectations ES1.06: Describe the factors that contribute to earth temperature gradients, and to wind speed and direction.

Pedagogical Issues
A high degree of integration is required of the students in these exercises. The most likely question to arise from students is "why does the low pressure system spin?"

Your explanation must be related to all of the elementary processes we have studied so far.

Science Issues
Two influences can be seen on these maps. The first is the general northward shift of the Hadley cells in January, and the southward shift in July. This is due largely to the tilt in the earth's axis. While the polar high pressure zones remain relatively stable over seasonal changes, the equatorial bands move markedly from winter to summer.

The second influence is the presence of land. The large continental land masses change temperature much more rapidly than the oceans. The heat loss from northern Asia, and from the Rocky Mountains of North America are prominent in January. The same regions are hot in July, with accompanying low pressures.

The pattern over the Pacific Ocean is quite different. Between 30° and 60° south latitude, there is very little land. The descending air of the Hadley cells appears to be much more consistent. This is especially true in July when the southern hemisphere receives less solar energy.

In either case, the rotation of the earth causes the high pressure and low pressure systems to spin in opposite directions.

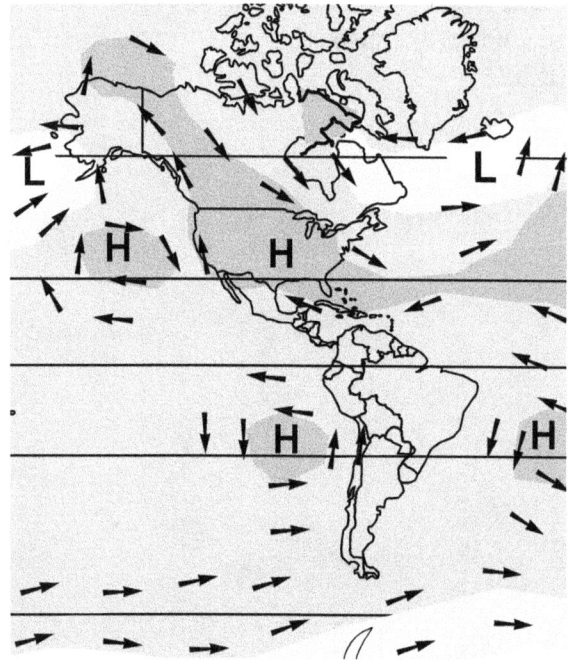

© Ross Lattner Publishing   44   www.rosslattner.ca

... on a Spinning Planet

Science and Pedagogy

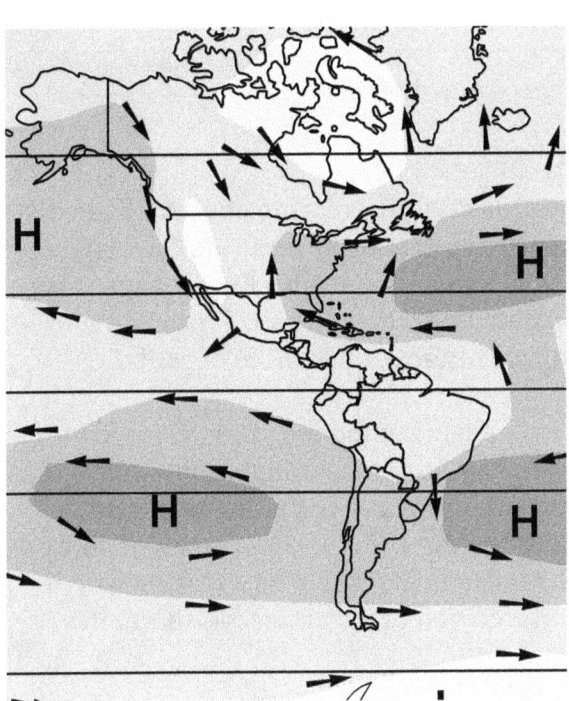

The Learning Activity

Students work through all of the exercises for each activity.

Two examples are shown.

Equipment, Preparation and Resources

Student exercises, pens, pencils *etc*.

Categories: **Assessment and Evaluation**
Knowledge and Understanding: Correctly determines the direction of rotating air masses
Thinking and Inquiry:
Communication:
Applications / Connections: Applies knowledge and skills to other maps

Explaining the Weather

10 Academic Science Teachers' Guide

Activity 3.5: Pressure Systems and Weather Maps

Learning Expectations ES1.06: Describe the factors that contribute to earth temperature gradients, and to wind speed and direction.

Pedagogical Issues

Reading a weather map, like reading any map, is an act of interpretation. Weather maps present a challenge beyond that of topographical maps. Reading a weather map is like interpreting a topographical map to figure out the geological history of that piece of earth.

Interpreting a weather map requires that the student comprehends the process at work in the atmosphere at the time that map was made. The most significant processes are the descent of cold, dry parcels of air by gravitational attraction, the ascent of warm, moist parcels, and the rotation of the air masses due to the spinning of the earth.

The weather map interpretation exercise in these pages concentrates upon those basic phenomena. The student must supply the information implied by the maps.

The maps in these exercises are simplified. Clouds, temperatures, humidity and precipitation are absent. Students should be able to infer the atmospheric conditions which would lead to the pressure patterns on the map.

Science Issues

The ability to bring a great many concepts together to provide a coherent explanation of weather is no small cultural achievement.

There are many additional influences at work on the weather around us. For example, the topography of the earth can cause funneling of air streams. Friction in the air disperses energy. These maps are simplified to ignore these more subtle, but nonetheless real, influences.

The pressure patterns are your evidence.

"What kinds of atmospheric conditions must exist that lead to the pressure patterns shown?"

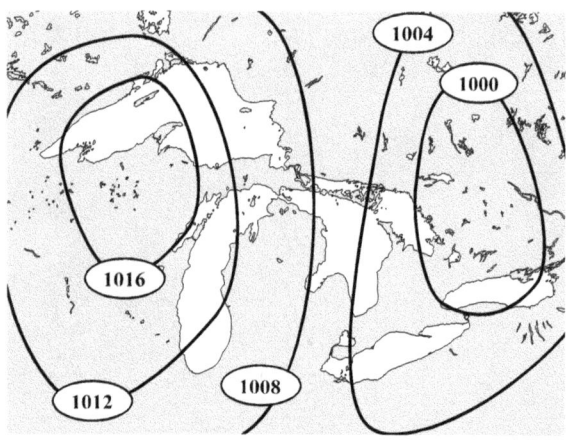

High pressure indicates a cool, dry, descending air mass. Low pressure indicates a warm, moist ascending air mass.

The air masses move in two different ways:

First, the air mass moves from high to low pressure, and second, each air mass rotates in response to the Coriolis effect.

... on a Spinning Planet

Science and Pedagogy

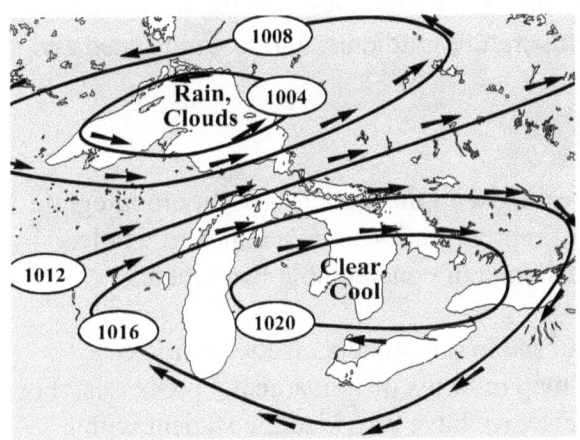

The Learning Activity
The students interpret each weather map according to the instructions. A sample is provided at left.

Equipment, Preparation and Resources
Students will use pens, pencils, *etc.* and the work sheets in the student exercise book.

Categories:
Knowledge and Understanding:
Thinking and Inquiry:
Communication:
Applications / Connections:

Assessment and Evaluation
Correctly interprets symbols on a weather map

Activity 3.6: Summer Front Systems in Southern Ontario

Learning Expectations ES1.06: Describe the factors that contribute to earth temperature gradients, and to wind speed and direction.

Pedagogical Issues

Once again, we are dealing with the student's ability to integrate a very large number of representations into a coherent whole. This is a cognitive challenge of considerable magnitude.

What we would like to see in our students is the ability to interpret the weather map in terms of fundamental processes. For example, in this exercise, we have provided the student with a conventional representation of a cold front, a warm front and an occluded front on the map in this exercise.

The student must undertake to interpret the front lines in terms of air masses, and then further interpret the air masses in terms of their temperature, humidity and density. Finally, the dynamic aspect of the situation must be interpreted.

This is a huge cognitive load for a young person. Yet it is, we believe, within reach as long as the student can undertake the interpretation in discrete stages. If we do not undertake this task with our students, then they are reduced to merely reciting descriptions or applying arbitrary rules without comprehending why fronts emerge and how they unfold.

Science Issues

The low pressure system that develops in the crotch where the warm and cold fronts meet is not the *cause* of the rainy weather. It is an *effect* of even more fundamental events and objects. The energy supplied by the sun, the presence of water vapour, the adiabatic cooling of ascending air parcels, and the changes of state undergone by the water all contribute to the developing low pressure system.

The behaviour of matter and energy remain the most important location for coherent scientific interpretations.

Sidebar:

Conventional Representation: a representation upon whose value the scientific community generally agrees.

The warm front / cold front representation is one of these. The representation bears resemblance to the events / objects in nature only to a small but important degree. The scientifically literate person must undertake considerable interpretation in order to make full sense of the whole.

... on a Spinning Planet

Science and Pedagogy

The Learning Activity

In this activity, the student is walked through a stepwise interpretation of a developing front system.

The student's task is to do two things:

1. Draw arrows to indicate the direction in which the developing low and high pressure systems rotate on the spinning planet.

2. Draw regions of cloud formation, paying particular attention to whether the clouds are likely to be cirrus, stratus or cumulus, and what elevations those clouds might reach.

The student's understanding of this weather system can be checked by having the student interpret a similar system in the quizzes which follow.

Further indications of student understanding may be found in the projects which the students undertake in the next section.

Equipment, Preparation and Resources
- photocopies of the student exercises
- pens, pencils *etc*.
- real weather maps for comparison

Categories:	Assessment and Evaluation
Knowledge and Understanding:	Must be assessed on another item from quiz or projects
Thinking and Inquiry:	Ability to interpret the consecutive maps of a weather system
Communication:	
Applications / Connections:	Ability to apply interpretation to other maps or systems

© Ross Lattner Publishing   www.rosslattner.ca

10 Academic Science Teachers' Guide — Explaining the Weather

Quiz 3.7: Spinning the Weather
Answer to quiz items

1 The high pressure zone (cool, dense sinking air) will have gone all the way around the earth in 24 h, and will have spread out over the surface of the earth. It will spin counter clockwise, as seen from space. 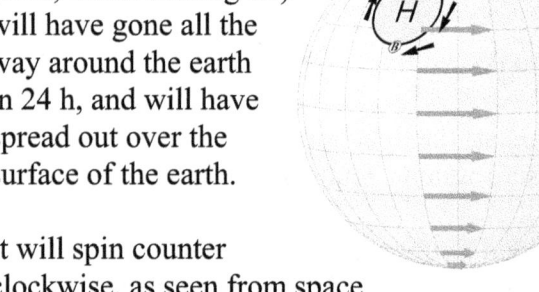	**2** The high pressure zone in the southern hemisphere must spin in the opposite direction from (1). 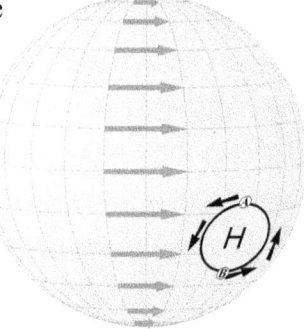
3 The low pressure (warm, moist, less dense) zone in the southern hemisphere will have ascended in 24 h and become smaller on the surface of the earth. The system will rotate more quickly than the earth, seen from the South Pole. 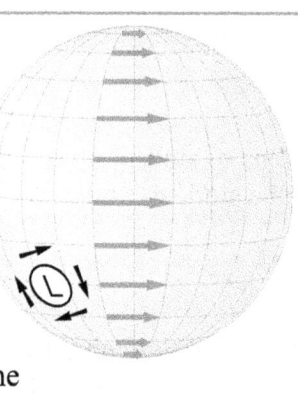	**4** The low pressure zone in the northern hemisphere will spin more quickly than the earth, as observed from the North Pole. 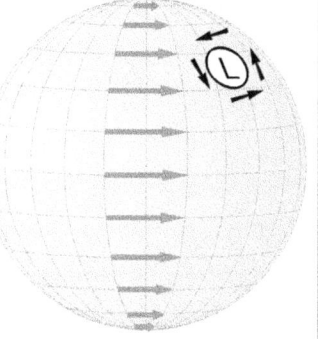
5 High pressure zones in general spin more slowly than the earth as they spread out. This makes them appear to be spinning in the direction opposite to the earth, in the respective hemispheres.	**6** Low pressure zones tend to be decreasing in size. They tend to spin more quickly as they contract, (like a spinning skater). The create winds that appear to revolve in the same direction as the earth, seen from the appropriate poles.
7 Rotational effects are greatest near the poles, and least near the equator.	**8** The sun heats the northern tropics in July, making them hot, humid, and less dense. These are the conditions for low pressure systems and rain. The opposite occurs in January.

Note: all of these models are very simple. In reality, low pressure zones would be rising, energy would be dissipating as radiated heat, and friction would change the motion of the rotating air.

9 The tropical air parcels contain 10 g of water, and the arctic air parcels contain less than 0.60 g of water (probably less than 0.30 g).

The low pressure zone will spin counter clockwise (in the same direction as the earth, but faster) and the high pressure zone will spin clockwise (in the opposite direction to the earth, as seen from the North pole. Near the front, the wind will blow from NE to SW, being fastest where the lines are closest. This is a very strong wind bringing bitterly cold air from the northeast.

10 Virtually all of the water in the tropical air parcel, i.e. 10.0 g, will freeze as the parcel rises.

Most of this will be happening at the point where the air parcels are rising most quickly, that is, at the developing low pressure zone over Green Bay Wisconsin.

11 The winds are now circulating counterclockwise around the developing low pressure zone

As 10 g of snow formed, the water vapour in the tropical air parcels released 28 300 J of heat energy, raising the temperature of each parcel by 28.3 °C. The air at 5 km in the vicinity of the low pressure zone is quite warm, and continues to rise strongly.

The heat energy released by the snow then increases the intensity of the low pressure cell, and this creates strong winds circulating around the low pressure cell.

12 Suppose we pick Toronto.

In diagram 9, the wind is warm, from the east, and increasing in strength.

By Diagram 10, intense clouds are beginning to form, and there is a change in the direction and / or intensity of the wind.

In diagram 11 there is a rapid decrease in temperature, intense snow fall, and the wind is strongly from the NW again. This is a blizzard!

By diagram 12, most of the snowfall has stopped, as the cold air has no water to make snow. The air is clear, very cold, and blue. The wind is swinging around to come directly from the North.

**10 Academic Science
Teachers' Guide**

Explaining the Weather

Possible Topics

*Famous Tornados
Famous Hurricanes
Famous Monsoons
Famous Blizzards
Famous Ice storms
Famous heat waves
Microclimates
Smog and Pollution
Global Warming
Greenhouse Effect
El Nino
Ice Ages
Weather on Mars
Weather on Jupiter
Weather on Venus
Dust Devils
Jet Stream
Hail storm*

Project 4.1: Extreme Weather Events

Learning Expectations ESV.01 - 02: Demonstrate an understanding of the factors affecting the fundamental processes of weather systems; investigate and analyze trends in local and global weather conditions to forecast local and global weather patterns.

Pedagogical Issues

This is the most challenging integration exercise in the unit. Students must use all of their knowledge, understanding, and inquiry skills to complete an investigation of extreme weather events.

A number of suggestions are provided at left. Add as many ideas as you can to this list. Some students will prefer to work with specific historical events, such as Hurricane Hazel.

Science Issues

The exercises so far have been focused upon student learning of the basic processes of atmospheric weather phenomena. Ocean currents, and upper atmosphere events, have been left out up to this point. We believe that the subject of ocean currents is certainly important for students to know, but that the dynamics of water flow are sufficiently similar to those of the atmosphere that a student could make a fruitful study of the ocean currents in this project.

Some other ideas:

the ice storm of 1998
nuclear fallout and weather
weather and important events, (e.g. WWII battles)
weather and early settlers in Canada
volcanic ash and its effects upon weather.

... on a Spinning Planet

Science and Pedagogy

Lightning
Trade winds
Recent storm in your area
Global ocean currents
Desert winds
Desertification
Weather in the Jungle
Weather in the Arctic
Weather in the Antarctic
Wind shear and Airplane Crashes
Downbursts and Airplane Crashes
Famous Forest Fires
High Altitude Sprites
High Altitude Jets
The Ozone Layer
Maritime Weather
Weather near Islands
The ice storm of 1998
Nuclear fallout and weather
Weather and important events, (e.g. WWII battles)
Weather and early settlers in Canada
Volcanic ash and its effects upon weather.

The Learning Activity

This activity can be undertaken over five days, either consecutively or distributed over the unit.

The general cycle is for students to:

1. Choose a topic and plan a resource search.

2. Conduct the resource search, and make up the bibliography.

3. Read the information collected, classify and organize it into some useful form.

4. Write the rough draft, and have it proofread by another student or parent.

5. Write the final draft and submit.

The short time line may look problematic, but the writing is actually not the greatest problem for most students. Choosing a topic, finding material, and studying it to find meaning is the most time consuming aspect of research.

You can take the pressure off by distributing these steps over the entire unit, or by placing a weekend between steps 3 and 4.

Equipment, Preparation and Resources

Library, computers, other sources of weather and related information

Categories: **Assessment and Evaluation**
Knowledge and Understanding: How readily does the student use his or her knowledge?
Thinking and Inquiry: Quality of the question, and the quality of the research
Communication: Quality of the writing and arguments
Applications / Connections: Application of basic concepts to new phenomena

Student Exercises
Explaining the Weather

Knowledge and Understanding

Three theories are emphasized in this unit. You are probably already familiar with the particle theory of matter. We will also use a specialized variation of the particle theory: the concept of an air parcel. Finally, we will use these together to develop a theory of atmospheric pressure.

You will learn how to calculate pressure, energy, temperature, mass and other quantities related to weather. In addition, you will learn how to draw specialized pictures of Earth's atmosphere, and to use these theoretical descriptions to explain and predict the weather around you.

Knowledge and understanding are probed at regular intervals in the Grade Ten Daily quizzes. Study these as you go through the exercises, so that you can do your best when they are assigned.

Inquiry and Thinking

We will use the PEOE cycle (Predict, Explain, Observe, Explain) for most labs and activities. You are expected to frame a question, provide your best prediction, and explain your thinking, using both sentences and diagrams. In other exercises, you will work with a variety of diagrams of air and water in the environment. You will use your diagrams to interpret and predict the weather.

At the end of the unit, you will be given a five day independent project. The project will demonstrate your ability to conduct your own investigation.

Communication

The quality of your arguments is the most important aspect of communication in this chapter. Your arguments consist of sentences, organized into paragraphs, and supported by diagrams or other representations.

Each sentence should be clear and to the point. Try to limit your sentences to just two concepts linked together to make a reasonable claim. If you need to relate more than two concepts, add a new sentence.

Applications, Connections and Extensions

Every exercise in this book is designed to support you as you learn appropriate theories and apply them to problems. In the labs, you demonstrate your understanding of a theory only by applying the theory. In the quizzes and projects, you are invited to make further connections and extensions of your learning.

Explaining the Weather

10 Academic Science Lab Manual

Introduction: Three Theories to Understand Our Weather

Suppose a scientist wishes to explain why a heavy dew appears on certain autumn mornings. The scientist must "tell the story" of the formation of the dew in terms of *things that are both simpler and more permanent* than the dew itself. The things that are simpler and more permanent than the dew are the particles of water that make up the dew. So we begin with one of the most important theories in science.

1. **The Particle Theory of Matter** consists of six simple statements which can help you explain things that happen all around you, and even help you to predict things you have never seen!

1.	The absence of matter is a pure vacuum.	The space between air particles is a vacuum, exactly the same as outer space.
2.	All matter is made of tiny particles.	Air is a solution of several gases: nitrogen, oxygen, water and carbon dioxide molecules, as well as argon atoms.
3.	All particles of one substance are identical.	
4.	The spaces between particles are small in solids and in liquids, and large in gases.	The distance between air particles is about 12 times larger than the particles themselves. In solids and liquids the space between the molecules is very small.
5.	All particles are attracted to each other by forces.	The attractive forces between the molecules of N_2, O_2, Ar, and CO_2 are so small as to be negligible. The forces between water particles can be quite strong.
6.	Particles are in constant motion.	Air particles average 450 m/s on a cold day. Imagine- that would be like being in a room full of golf balls traveling faster than rifle bullets!

2. **The Concept of "Air Parcels"** The whole atmosphere is far too large to imagine all at once. A "parcel" of air is an imaginary little packet of air. If we choose its properties carefully, we can use the idea of an air parcel to explain a great many things. Let's choose to use a parcel of dry air, exactly 1.0 kg in mass. Such a parcel, in your classroom, would have a volume 1 m × 1 m × 78 cm high.

    1. The *mass* of an air parcel is equal to the total mass of all of its particles.

    2. The *temperature* of an air parcel is a measure of the speed of the molecules.

    3. The *pressure* of an air parcel depends upon the masses of the parcels above it.

    4. The *volume* of a parcel gets larger with increased temperature.

    5. The *volume* of a parcel gets smaller as air pressure is increased.

    6. The *relative density* of an air parcel is equal to its *mass ÷ volume*.   D = m/V

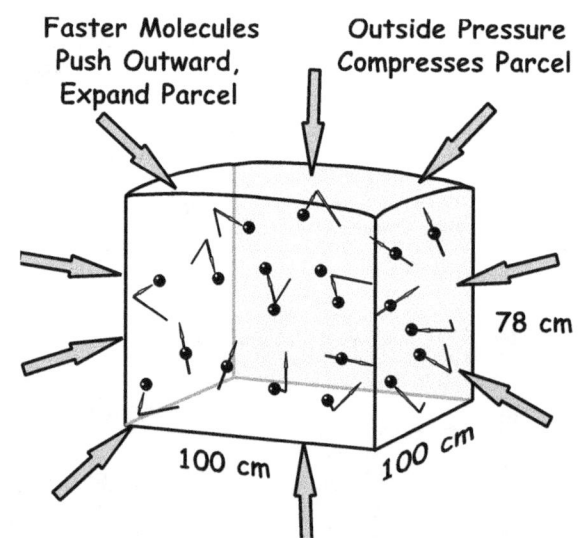

Introduction

Name:
Date:

3. A Theory of Atmospheric Pressure

The atmosphere over your desk can be thought of as a stack of air parcels, about 100 km high. There are only two fundamental forces at work in the atmosphere above you.

1. The *gravitational force* attracts all matter down toward the earth. The bottom parcel is squeezed by all of the parcels on top. This force is large!

2. The *electrical force* is always carried by light. Sunlight provides the energy that keeps air particles moving at a rate of about 450 m/s near earth. At that speed, they push each other apart with their collisions.

The only thing that keeps the earth's gravity from crushing the atmosphere completely to the ground is the kinetic energy of the air particles themselves.

1. **All air molecules have mass, so they experience a downward force of gravity.**

2. **All of the air molecules above a parcel of air press downward on that parcel.**

3. **Kinetic energy from sunlight makes molecules move at about 450 m/s near the ground.**

4. **Each parcel of air supports all of the molecules above it.**

5. **Warmer (faster) particles expand a parcel; cooler (slower) particles are squeezed together.**

6. **At the same pressure, cool air is more dense than warm air.**

Write each of the six points of the theory on the diagram at right. Write words and draw arrows to help you remember these six points.

In these exercises, the question must be answered in *complete sentences*. One sentence is one thought. A single word is simply not enough.

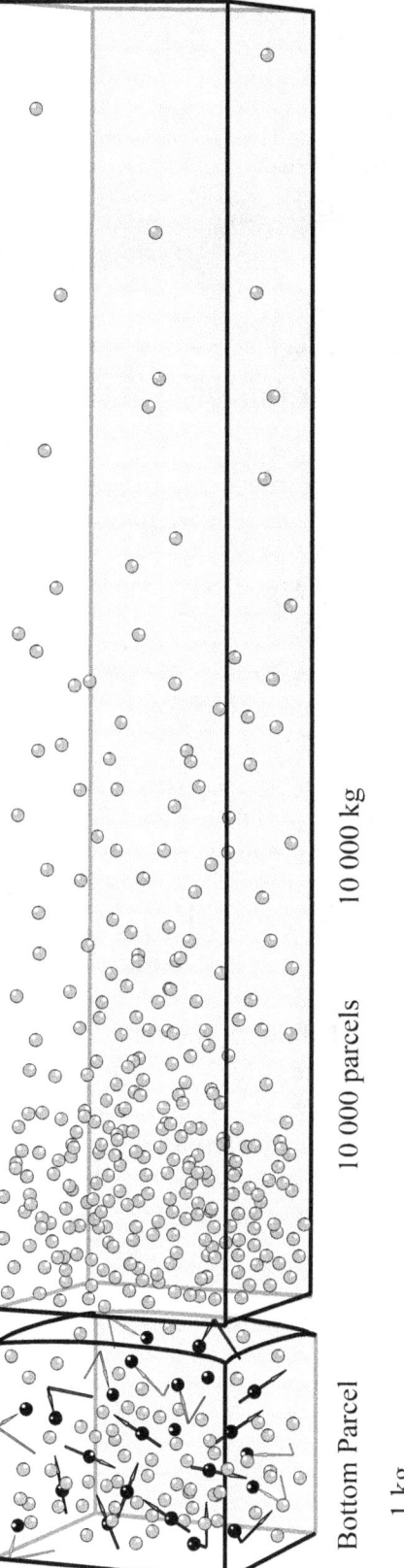

© Ross Lattner Publishing   www.rosslattner.ca

Explaining the Weather

10 Academic Science Lab Manual

Activity 1.1: Your Daily Weather Log

Do you Remember? Define these terms to the best of your ability, without looking them up.

1. Pressure
2. Temperature
3. Humidity
4. Density
5. Composition
6. Wind Speed

Change these definitions as you learn more about these concepts in the following pages.

What's The Question? You have probably seen the weather forecasts either on television or a newspaper. Perhaps you have noticed that they are sometimes incorrect. *How good are the weather predictions from your weather source?* Use only one weather source, either a newspaper or TV.

What Are We Doing?

1. **Predict** the fraction of correct 3-day forecasts in your newspaper or television station. Is it 90% correct? 50% correct? Or some other %?
2. **Explain** why you made your choice. Use as many "weather words" as you can.
3. a) **Observe** one weather source for four weeks. Make records of both the prediction for the third day ahead, and the weather today as it actually happened.
   b) Look up a satellite weather map on a web site each day at the same time. Print out each weather map, and keep them with your logs.
   c) Calculate the % that the weather forecaster was able to correctly predict 3 days in advance.
4. **Explain** any differences between what you predicted and what you observed. Does your weather source make better or worse predictions than you thought?

What Are We Thinking About?

We are looking for connections between different aspects of the local weather.

- What influences appear to make *pressure* vary?
- What influences appear to make *temperature* vary?
- Do any factors always appear to arrive together? What factors are they?

Questions For Later...

1. What seems to be the most difficult thing for weather forecasters to predict?

2. Do you think that weather forecasting is more difficult in summer, winter, spring or fall? Why?

Energy and the Weather

Name:
Date:

Focus Question: Write the question that you are trying to answer.

1 **Predict** the % accuracy your weather source provides when it forecasts the weather

2 **Explain** why you made your choice. Use as many "weather words" as you can.

3 **Observe** and make records. Your teacher may provide you with a set of forms to assist you.

4 **Explain** any differences between what you predicted and what you observed.

10 Academic Science Lab Manual

Explaining the Weather

Activity 1.2: Conduction, Convection and Radiation
Do you Remember? Write the 6 propositions of the Particle Theory of Matter.

1. _____
2. _____
3. _____
4. _____
5. _____
6. _____

What's The Question? Thermal energy is all about particles. Particles in the atmosphere can exchange energy among themselves and with the environment in three ways: radiation, convection and conduction. *How do these three means of heat transfer work in our weather environment?*

What Are We Thinking About?

1. **Conduction:** collision between individual particles. When a fast particle (A) collides with a slow particle (B), energy is transferred from the fast to the slow particle (C). Thermal energy is transferred from left to right.

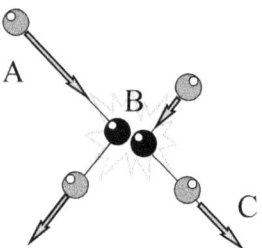

2. **Convection:** In a cool air mass, particles are close together. In a warm air mass, they move farther apart. The cooler mass is more dense, and is more strongly attracted to earth. Gravity causes the denser mass to sink (A), forcing the warmer mass to rise (B). Heat is carried upward by the rising warm air.

 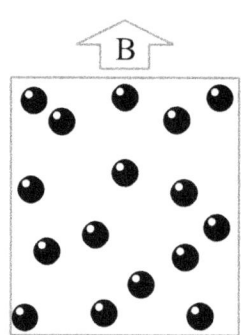

3. **Radiation:** (A) a particle with energy gives up some of that energy in the form of radiation. In the process, the particle loses energy, and slows down. (B) The infrared and visible light carries energy from one particle to another. (C) another particle absorbs the radiation, and gains the energy. Thermal energy is transferred from left to right.

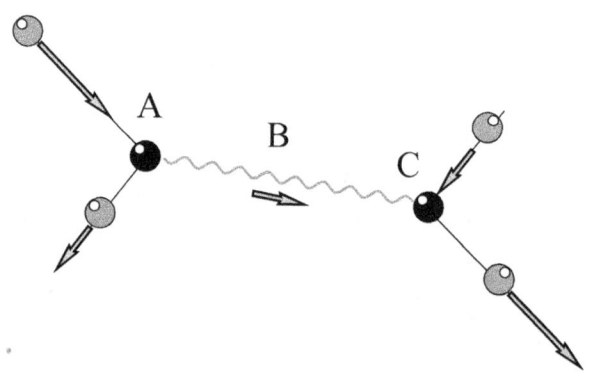

Energy and the Weather

Name:
Date:

What Are We Doing? Label the diagram below to show each of the following:

1. a) warm air
   b) cool air
   c) warm solid
   d) cool solid

2. the direction of air flow (four arrows)

3. one example of each kind of heat transfer.
   a) conduction
   b) convection
   c) radiation

4. Add radiation waves to show heat transfer from:

   a. sun to solid particles
   c. air to air particles
   b. sun to air particles
   d. solid to air particles.

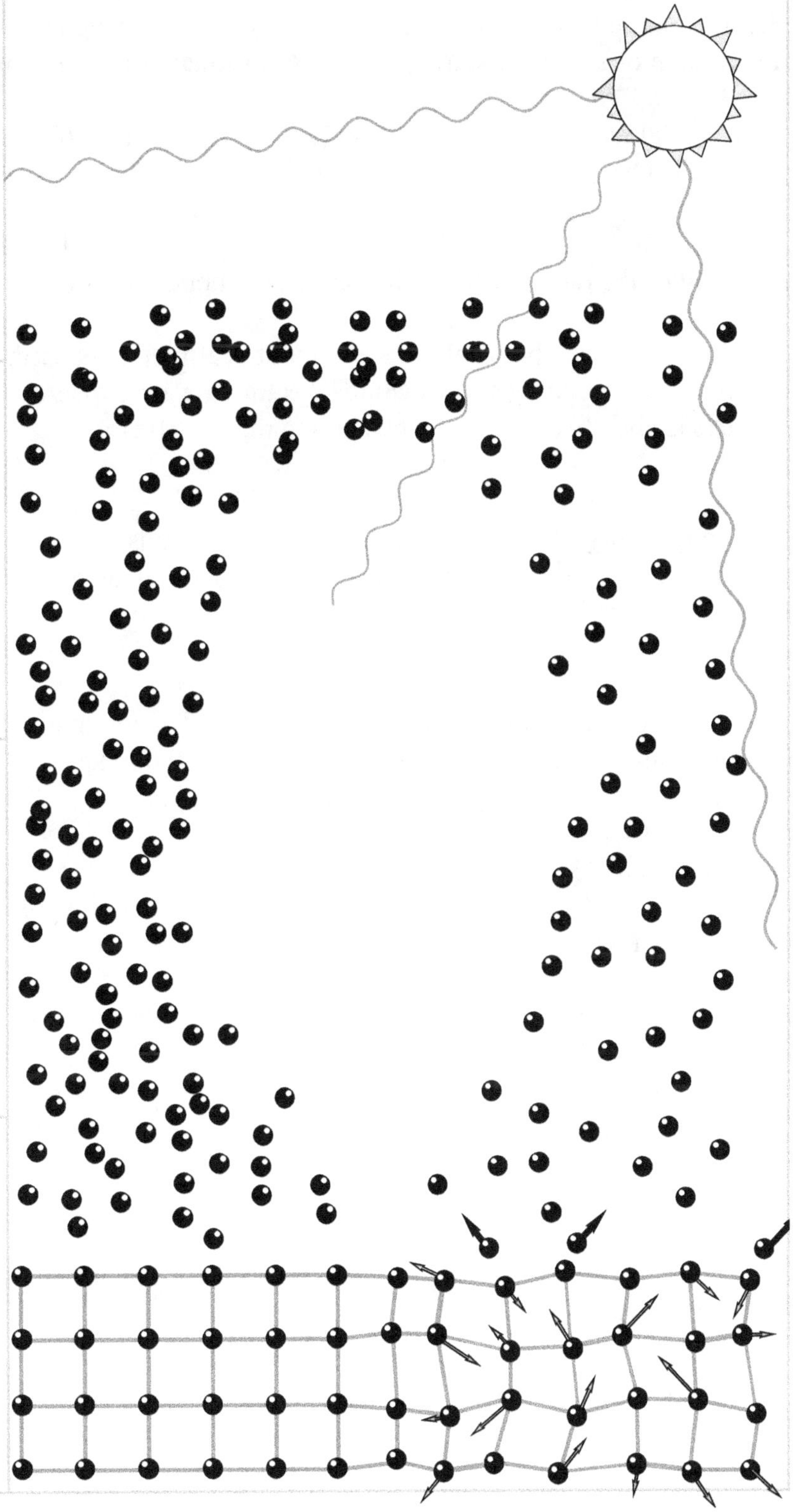

© Ross Lattner Publishing   www.rosslattner.ca

Energy and the Weather

Name:
Date:

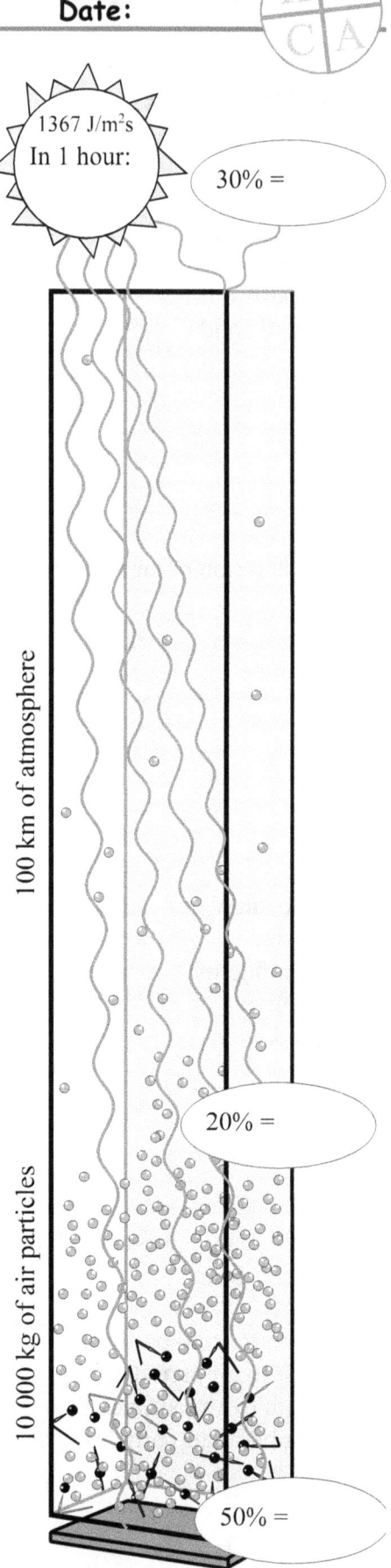

What's The Question? All of the energy that drives our weather comes from the sun in the form of electromagnetic radiation (light). *What happens to all of the sun's radiant energy as it*

What Are We Thinking About? Consider a 1m × 1m patch of sky at noon on a clear, sunny day. The light passes through a 100 km column of air, finally striking a 1m × 1m square of pavement.

1. The total radiation reaching the upper atmosphere is 1367 J upon each square metre, every second.

2. Of all of the radiant energy that strikes the upper atmosphere,

   30% is reflected back into space without heating planet earth.
   20% is absorbed by air and clouds, heating the atmosphere
   50% is absorbed by pavement, heating the top 10 cm.

   The **Q**uantity of heat transferred to a body depends upon the **m**ass of the body, the heat **c**apacity of the material, and the **T**emperature change. $Q = mc\Delta T$

3. The total mass of the atmosphere in the 1m × 1m column is ten thousand kilograms. That's right. m_{air} = 10 000 kg. The total mass of the pavement, m_{pave} = 500 kg.

4. The *specific heat capacity* is the quantity of heat required to heat one kg of a substance through one Celsius degree of air

   c_{air} = 1000 J/kg • °C (1000 joules will heat 1 kg air by 1°C)
   c_{pave} = 900 J/kg • °C
   c_{water} = 4180 J/kg • °C.

What Are We Doing?

Use the information from this page to answer all of the questions on the following page.

© Ross Lattner Publishing   62   www.rosslattner.ca

Energy and the Weather

Name:
Date:

1. The sunlight that reaches our upper atmosphere is 1367 joules of solar energy each square meter every second. How much solar energy reaches earth, in joules per square meter per *hour*?

2. Use the percentages from previous page calculate how many joules of solar energy in one hour are:

   a. reflected back into space (Quantity of heat reflected = Q_r)

   b. absorbed by the air in one hour Q_{air}.

   c. absorbed by the pavement in one hour Q_{pave}

3. If we rearrange the formula $Q = mc\Delta T$, we get an equation for ΔT

   $$\Delta T_{(air)} = \frac{Q_{(air)}}{m_{(air)} c_{(air)}}$$

   Use the new equation to find ΔT_{air}

4. Use the new formula to find ΔT_{pave}. Hint: you will need to use the appropriate quantities for heat of the pavement, mass of pavement, and specific heat capacity for pavement.

Does the sun's radiation heat the air directly? Or is the air heated by some other means? Explain your thinking.

10 Academic Science Lab Manual

Explaining the Weather

Activity 1.3: Energy and Changes of State of Water

What's The Question? Water is prominent in almost every environment. Water molecules have a strong electrical attraction for each other, so it takes a considerable amount of energy to force them apart. *How much energy is absorbed or emitted as water changes form?*

What Are We Doing? Answer each of the following questions. Show all of your work on a separate page.

What Are We Thinking About? Each diagram indicates a change of state of water, and the energy lost or gained per gram.

1.a How much energy is needed to melt 10 kg of ice?

1.b How many grams of ice could be melted with 1000 J of energy?

1.c If all of the solar energy that fell on the 1 m² slab of asphalt pavement was used to melt ice, what mass of ice could be melted in one hour?

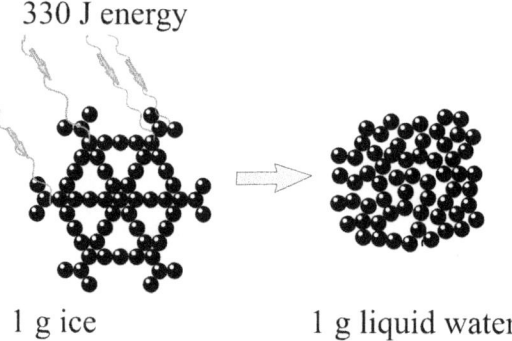

330 J energy

1 g ice → 1 g liquid water

2.a How much energy is needed to vaporize 250 g of water in a drinking glass?

2.b How many grams of water could be vapourized by a 100 W light bulb in one hour (360 000 J)?

2.c If all of the solar energy falling on a 1 m² slab of pavement vapourized the water, how many grams of water could be dried up in one hour?

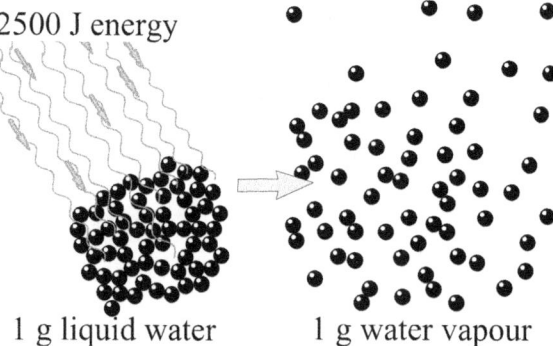

2500 J energy

1 g liquid water → 1 g water vapour

On a cold day, ice crystals can dry up without first turning to water (*sublimation*). It takes the same amount of energy as it would to both melt and vaporize the water.

3.a On a cold, sunny day, how many grams of snow could be dried off a 1 m² slab of sidewalk in 1 h ? (Assume 2.46 MJ of solar energy, p 63.)

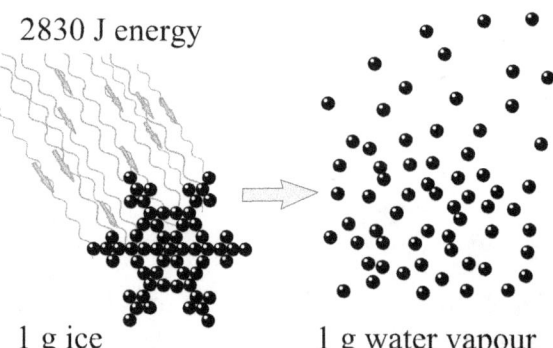

2830 J energy

1 g ice → 1 g water vapour

Energy and the Weather

Name:
Date:

4.a Imagine a 10 km × 10 km square of water in Lake Erie, freezing 2 cm deep one night. How much heat energy would be released into the environment? (Hint: one cubic metre of water is 1 000 000 g)

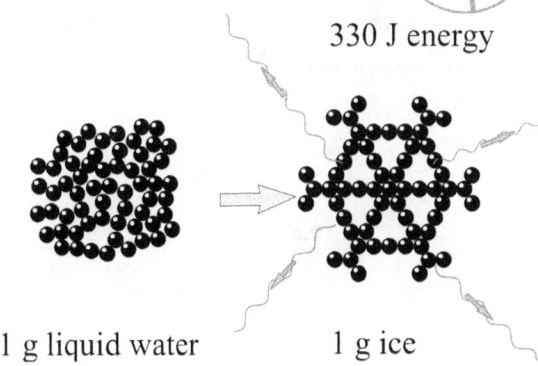

330 J energy

1 g liquid water   1 g ice

5.a How much energy would be released if 12 g of water condensed on the outside of a chilled glass of water?

5.b How many grams of water would have to condense from vapour in order to release 10 000 J?

5.c If 1 mm of dew condensed on a 1 m^2 patch of lawn, it would amount to 1000 g of water. How much heat energy would be released?

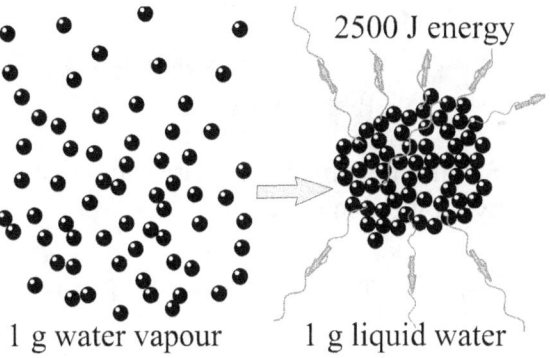

2500 J energy

1 g water vapour   1 g liquid water

Water vapour can freeze directly into ice in the process called *sublimation*.

6.a 3.0 cm of snowfall on a 1 m^2 patch of sidewalk represents about 3000 g of snow. How much heat energy was released as the snow formed?

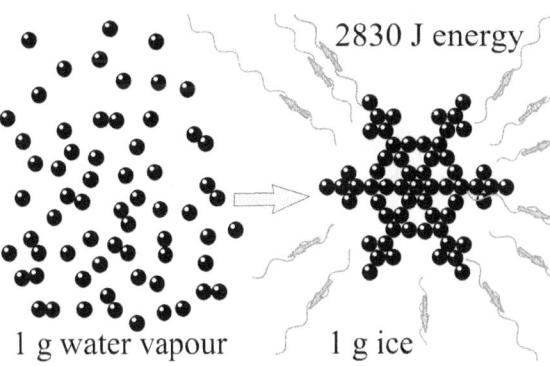

2830 J energy

1 g water vapour   1 g ice

© Ross Lattner Publishing   www.rosslattner.ca

Explaining the Weather

10 Academic Science
Lab Manual

Lab 1.4: The Concept of Atmospheric Pressure
Do you Remember? Write the 6 propositions of the Particle Theory.

1. _____
2. _____
3. _____
4. _____
5. _____
6. _____

What's The Question? You have frequently heard about air pressure. Perhaps you already know that the average air pressure on earth is about 101.3 kiloPascals (kPa), or 1013 millibars (mb).

What is air pressure? Where does it come from? How does it vary within the atmosphere?

What Are We Doing?
The size of the atmospheric pressure is difficult to experience.

1. **Predict** how much pressure can you exert by blowing into a tube?

2. **Explain** why you believe your prediction.

3. **Observe** your own results when you blow into the open end of a water manometer.

4. Your lungs and muscles are not in contact with the water in the manometer. **Explain** how you can hold up a water column by blowing into the manometer.

What Are We Thinking About?
A manometer is a simple device to measure air pressure. It consists of a simple U-shaped tube that is open at both ends. When you blow into the lower end, you can push water around the U.

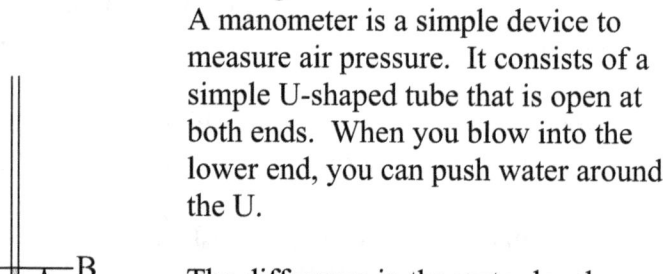

The difference in the water levels can be used to measure how much greater the air pressure is at A than it is at B.

Pressure = 9.8× h
P = 9.8 h

Pressure, P   kPa
Height, h   m

Try these for practice...
1. The height of the water column in the diagram above is 0.11 m. What is the pressure of the water?

2. The air pressure on the water at **B** is 101.3 kPa. What is the air pressure on the water at **A**?

3. What pressure must your lungs and muscles be exerting on the air?

4. How many times harder would you have to blow in order to exert 101 kPa at A?

© Ross Lattner Publishing   www.rosslattner.ca

Energy and the Weather

Name:
Date:

Focus Question: Write the question that you are trying to answer.

1 Predict the difference in height that your breath could push the water up the manometer tube.

2 Explain your prediction, using full sentences. Try to refer to any experiences you may have had with air pressure.

3 Observe the actual difference in height that you are able to push the water using your own lungs and muscles.

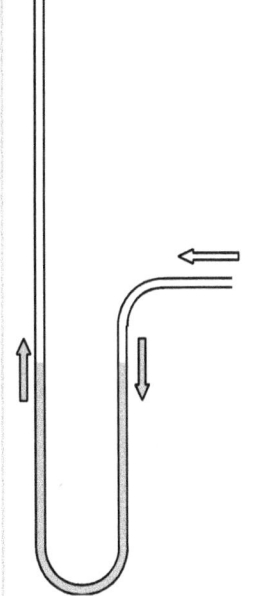

4 Explain any differences between your prediction and your observations. What have you learned?

10 Academic Science
Lab Manual

Explaining the Weather

Activity 1.5: Atmospheric Pressure
Do you Remember? Write the 6 propositions of the Theory of Atmospheric Pressure.

1. _____
2. _____
3. _____
4. _____
5. _____
6. _____

What's The Question? What is the pressure structure of the atmosphere?

What Are We Doing? Draw pictures and write labels to put *all* of the pressure data from the following tables onto the graph of altitude on the opposite page. You live at the bottom of an ocean of air!

What Are We Thinking About?
Average atmospheric pressure at sea level is 101.3 kPa. Most weather sources use 1013 mb. See how they are related?

- Sea Level is considered to be 0 m
- Lake Ontario lies at an altitude of 80 m
- Toronto is at about 100 m
- Small planes fly at an altitude of 1 km
- The CN tower is 300 m above L. Ontario
- Commercial jets fly at an altitude of 5 km
- The Rocky Mountains reach 4 - 5 km
- Military jets often reach heights of 10 km
- The space shuttle orbits at 200 km
- Weather satellites orbit at 500 km
- TV satellites orbit at 18 000 km
- At 1 km, 13% of the atmosphere is below you
- 51% of the atmosphere lies below 5 km
- 76% of the atmosphere lies below 10 km

- 94% of the atmosphere lies below 20 km
- When you reach 50 km, you are above 99% of the air molecules in the atmosphere! The sky is black, you can see the stars, you're almost "out in space"
- At 100 km, 99.9999% of the atmosphere is below you
- *Aurora Borealis* occurs from 100 - 400 km
- At 0 m, atmospheric pressure is 1013 mb
- 100 m, atmospheric pressure is 998 mb
- At 200 m, P = 984 mb
- At 300 m, P = 970 mb
- At 400 m, P = 956 mb
- When you reach 500 m, P = 943 mb
- At 1000 m, P = 877 mb
- At 5 km, atmospheric pressure is 493 mb
- At 10 km, P is only 240 mb
- At the stratosphere, 15 km, P = 112 mb
- At the edge of space, 50 km, P = 0.76 mb
- Most of the weather that you experience happens below 12 km and so is most of the water and the pollution
- A 1.0 kg parcel of air, 1 m × 1m at the bottom, is 78 cm tall at ground, 100 cm tall at 2 km, and 200 cm tall at 8 km

The graph on the opposite page is a *logarithmic scale*. Note that small altitudes are spread out, but larger altitudes are squeezed together. Scientists use *log scales* on graphs to display data that includes both very large and very small numbers. The altitudes in the earth's atmosphere go from 0.1 km all the way out to 1000 km. You couldn't see all of that information on a standard linear scale.

© Ross Lattner Publishing   www.rosslattner.ca

Energy and the Weather

Name:
Date:

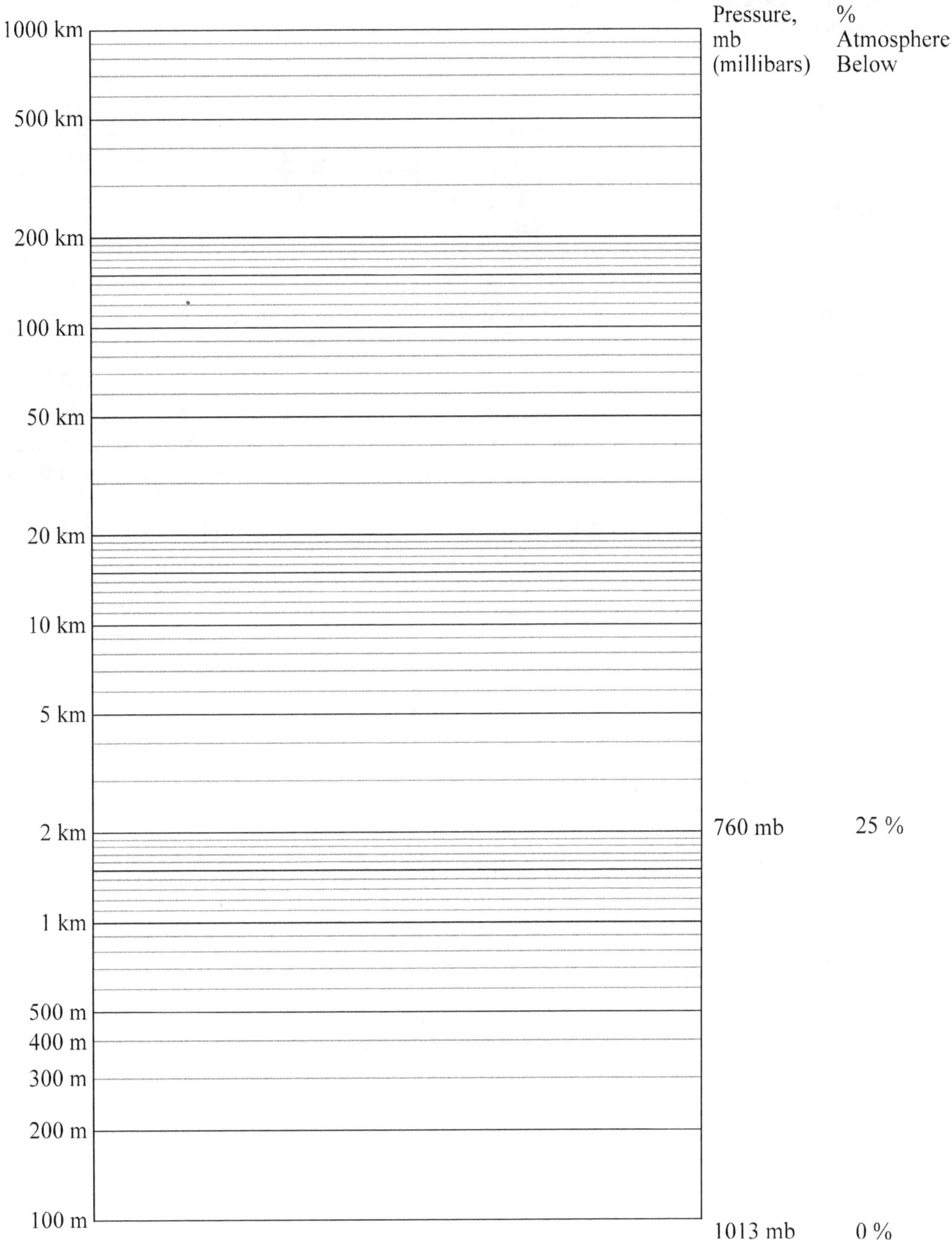

Altitude	Pressure, mb (millibars)	% Atmosphere Below
1000 km		
500 km		
200 km		
100 km		
50 km		
20 km		
10 km		
5 km		
2 km	760 mb	25 %
1 km		
500 m		
400 m		
300 m		
200 m		
100 m	1013 mb	0 %

Level _____

© Ross Lattner Publishing   69   www.rosslattner.ca

Explaining the Weather

10 Academic Science Lab Manual

Activity 1.6: Temperature in the Atmosphere
Do you Remember? Write the 6 propositions of the Ideal of Air Parcels.

1. _____
2. _____
3. _____
4. _____
5. _____
6. _____

What's The Question? We all know how temperatures change here on earth from day to night, and from summer to winter, but... *How does atomospheric temperature vary from earth to space?*

What Are We Doing?
Draw pictures and write labels to put *all* of the temperature data from this page onto the graph of altitude on the opposite page.

- The average temperature at sea level is 15°C
- Most of our weather happens in the *troposphere*, 0 - 12 km
- Temperature decreases steadily to –60°C at 12 km
- In still air, the temperature is about 6.3 °C cooler for each 1 km increase in altitude
- Above 12 km, the temperature begins to increase again
- Ozone absorbs UV light energy, and transforms it into heat
- Almost all of the *ozone layer* lies in within the *stratosphere* at 12 - 50 km
- The top of the stratosphere is the edge of space
- The *mesosphere* (*middle* atmo*sphere*), 50 - 100 km, is very thin. Almost all ultraviolet passes completely through it.
- Ultraviolet light from the sun is absorbed by oxygen atoms in the *thermosphere*, 100km and above, making it very hot.
- Ultraviolet light that makes it through the thermosphere is absorbed by the ozone in the stratosphere.
- The thermosphere is 500°C at 200 km, and 700°C at 300 km!
- Even though the thermosphere is very hot (high speed particles), there are so very few particles that your hand would freeze in minutes out there.
- Oxygen molecules O_2 are broken into single oxygen atoms by the solar radiation above 100 km. Free oxygen atoms are known to corrode satellites very quickly.

What Are We Thinking About?
- Temperature is a measure of the average kinetic energy of the air molecules.
- At 0°C, the average speed of air molecules is ≈ 447 m/s

Altitude, km	Temperature, °C
0	15
2	3
4	-10
6	-23
8	-35
10	-48
12	-60
50	0
80	-90
100	20
150	150

- These figures represent an *average* temperature near the equator. **Real temperatures vary greatly from winter to summer, and from the equator to the poles.** The lowest 12 km of atmosphere are the most changeable.

© Ross Lattner Publishing   www.rosslattner.ca

Energy and the Weather

Name:
Date:

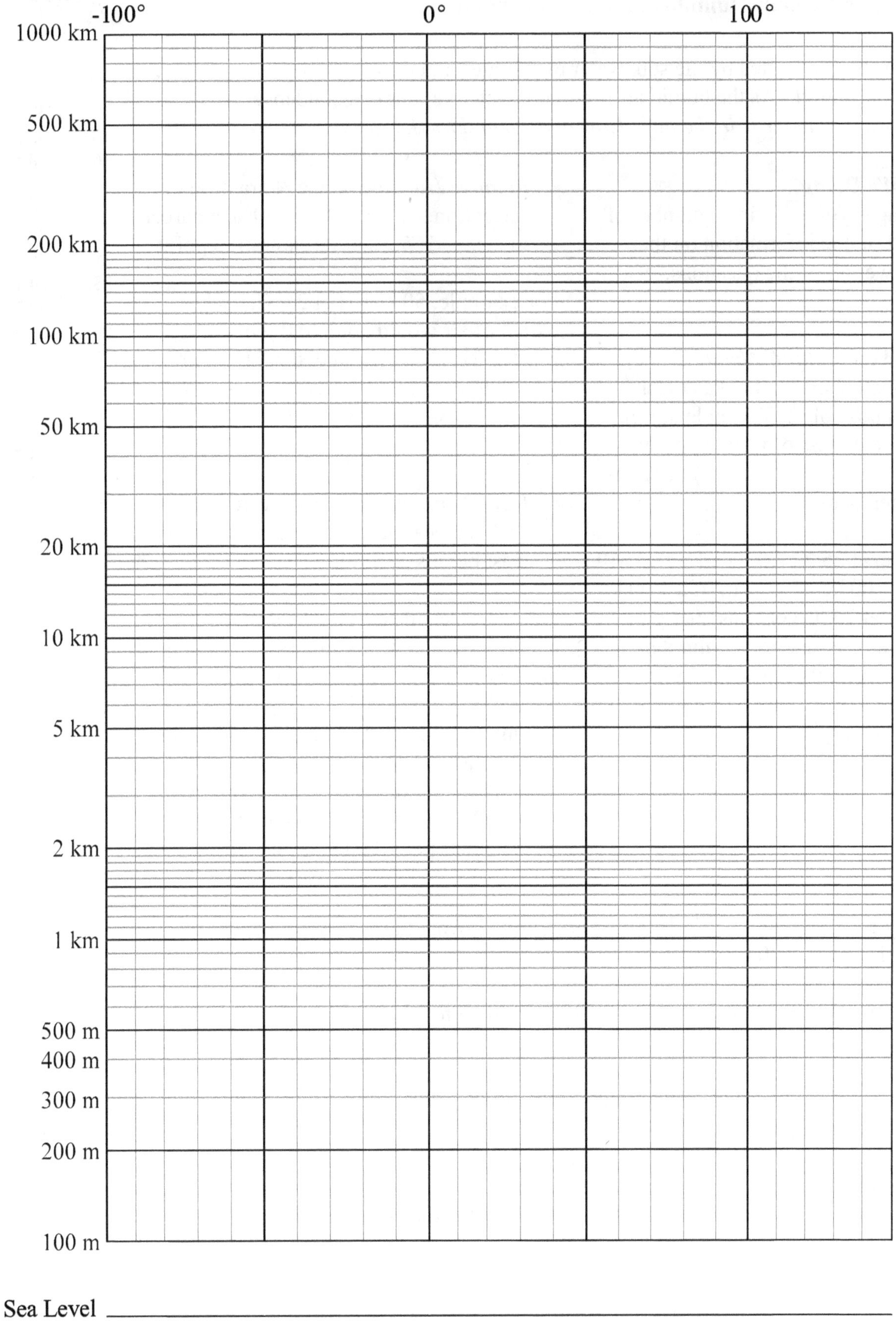

-100°   0°   100°

1000 km
500 km
200 km
100 km
50 km
20 km
10 km
5 km
2 km
1 km
500 m
400 m
300 m
200 m
100 m

Sea Level _____

© Ross Lattner Publishing   71   www.rosslattner.ca

10 Academic Science Lab Manual
Explaining the Weather

Lab 1.7: Water Vapour, Humidity and the Dew Point

As air and water are heated by the sun, water evaporates and dissolves in the dry air, making the air humid. As the humid air is cooled, water vapour can condense. *How much water vapour is in the air? How and when does dew form?*

What Are We Doing?
On these pages you will find a number of problems. Use the information on this page to answer all of the questions. Show your work.

What Are We Thinking About?
- *Saturation* means "to hold as much solute as possible". The table at right tells you the mass of water that will dissolve in a parcel (1.00 kg) of dry air at each temperature.

For example, at 25°C, 20.1 g of water will dissolve in 1.00 kg of dry air. If a 1.00 kg parcel of air at 25°C contains 20.1 g of water vapour, it is said to be saturated.

A How many grams of water vapour will 1.00 kg of dry air hold on a cold (–5°C) day in February?

- *Relative Humidity* is the % fraction of saturation. If the RH% is 50%, a parcel of air holds 50% of its saturation point.

For example, 50 % RH at 25°C, each parcel of air would hold 10.0 g of water.

B One 22°C day in June, the RH is 30%. What mass of water will dissolve in a 1.00 kg parcel of dry air?

- *Dew Point*: the temperature at which air becomes saturated, and dew, fog, or frost begins to form.

For example, air that is 50% RH at 25°C would have a dew point of 14°C. That is the T at which a 1.00 kg parcel of air would be saturated with 10.0 g of water.

C One evening in August, 25°C and 60% humidity, a clear sky prevails and the temperature falls to 11°C. Will dew form? If so, what mass of dew will condense per 1.00 kg air parcel?

T°C	g H_2O
40	48.4
38	43.5
36	38.8
34	34.5
32	30.7
30	27.2
29	25.6
28	24.1
27	22.7
26	21.3
25	20.1
24	18.9
23	17.8
22	16.7
21	15.7
20	14.7
19	13.8
18	13.0
17	12.1
16	11.3
15	10.7
14	10.0
13	9.3
12	8.7
11	8.1
10	7.6
9	7.1
8	6.6
7	6.2
6	5.8
5	5.4
4	5.0
3	4.7
2	4.4
1	4.0
0	3.8
-1	3.5
-2	3.3
-3	3.0
-4	2.8
-5	2.5
-6	2.4
-8	2.1
-10	1.7
-12	1.5
-14	1.3
-16	0.9
-18	0.8
-20	0.6

1. It is very cold outside, -20°C. If the air is saturated, what mass of water vapour is dissolved in each 1.00 kg parcel of air?

2. If the air in (1) comes into your house, and is warmed to 22°C without having any water vapour added, what is the humidity inside your house?

3. One October day, T = 18°C and humidity 40%, the temperature falls suddenly overnight to -5°C. Will frost form?

4. It's a hot summer day, 32°C and 40% humidity. You get a bottle of water from inside the house. No dew forms on the bottle. What is the coldest temperature the bottle could be?

Energy and the Weather

Name:
Date:

5. a. On a fine day in May, the air is 28°C and 40% humidity. What mass of water would be dissolved in each 1.00 kg parcel of air?
   b. If you took all of the air in a block 1 km × 1 km × 780 m, how many parcels of air would you have? (Recall that a single parcel of dry air is about 100 × 100 × 78 cm)
   c. If all of that nice May air rises to a height of 2 km, where its temperature becomes 6°C. What mass of water condenses out as a cloud of water droplets?

Focus Question: *What is the humidity in the room right now?* Given one thermometer, a metal can, water and ice and paper towels, make up a way to find the dew point of the air in the room. Then calculate the RH% (Relative humidity %) in the room.

1 Design your experiment, and *predict* the RH% in the room.	2 *Explain* how your experiment works, and give reasons for your prediction.
3 *Observe* the results, and make records of them. Describe any changes you had to make as well.	4 *Explain* your results.

© Ross Lattner Publishing

10 Academic Science Lab Manual

Explaining the Weather

Activity 1.8: Relative Density of an Air Parcel
Do you Remember? Write 6 statements that describe the Idea of an Air Parcel.

1. _____
2. _____
3. _____
4. _____
5. _____
6. _____

What's The Question? Be patient! We are getting close to understanding the earth's weather!

What is the relative density of air? Can we predict whether the parcel will rise, sink, or rest?

What Are We Thinking About?

- The density of a parcel of air is its *mass per volume*.

  $$D = \frac{m}{V}$$

  Since we have decided that an air parcel is 1.00 kg of dry air, only the volume of the air parcel changes.

- Increasing the temperature by 1 °C will make the air parcel taller by ≈3 mm. Cooling does the reverse.

- Each gram of water picked up by 1.00 kg of dry air has the same effect as increasing the height of the air parcel by 0.50 mm.

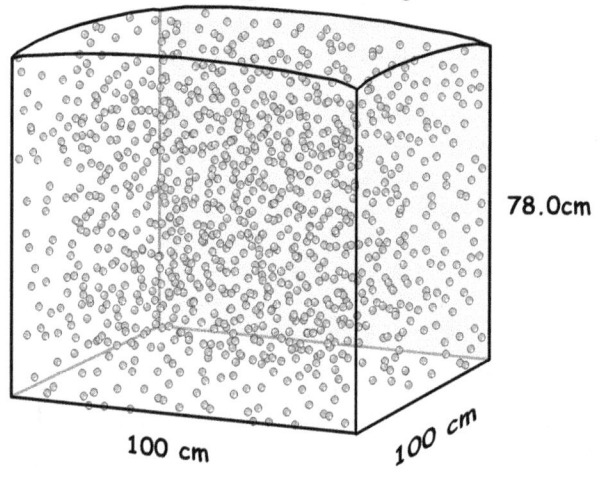

Parcel of air at 0°C and 1013 mbar

Add 2.75 mm to height for each degree warmer
Add 0.50 mm to height for each gram of water

78.0cm
100 cm
100 cm

Mass of air particles: 1.00 kg

Questions For Later...

1. Consider a teeter totter with a 100 kg person and a 50 kg person. Who sinks, who rises? Which person "causes" the motion? Explain your thinking.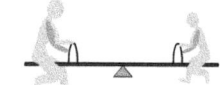

2. Now consider two air parcels, one more dense, the other less dense. Which parcel rises, which one sinks? What "causes" the motion? Explain!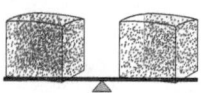

© Ross Lattner Publishing   www.rosslattner.ca

Energy and the Weather

Name:
Date:

In each problem, there are two air parcels: yours, and the surroundings. **Calculate** the relative density of two air parcels. *Show all of your work!* Decide whether your air parcel will rise, sink, or rest relative to the surroundings. **Explain** your thinking.

1. Your air-parcel is at 28°C. The surrounding air parcels are at 16°C. Calculate the height and volume of each parcel. Calculate both densities. Which one sinks? Which one rises?

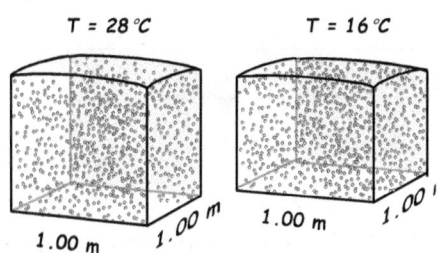

2. Your air parcel is at 9°C, and is floating just below a cloud. Your surroundings are at 15°C.
Calculate densities and decide whether you will sink or rise.

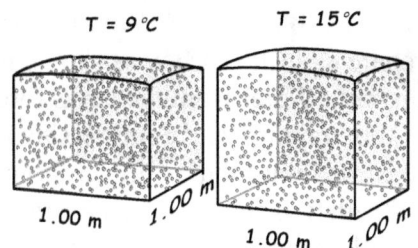

3. Wow! You have risen to 2000 m, where the surrounding air parcels are at 4°C. Your air parcel is at 13°C, and 90% RH..
Find the densities of your air parcel and that of the surroundings. Will you rise, sink or rest?

4. You have descended to region that is 25°C and 0% RH. Your air parcel is at 24°C and 80% RH. Find both densities, and predict movement.

© Ross Lattner Publishing   75   www.rosslattner.ca

10 Academic Science
Lab Manual

Explaining the Weather

Lab 1.9: Formation of a Cloud
Do you Remember? Write the six propositions of the Theory of Atmospheric Pressure.

1. _____
2. _____
3. _____
4. _____
5. _____
6. _____

What's The Question? Rising air expands as at reaches ever smaller pressures on the way up. Likewise, sinking air is compressed as it descends, as more and more of the earth's atmosphere is above it. *When an air parcel expands, does its temperature increase, decrease, or stay the same?*

What Are We Doing? Your teacher will prepare a plastic carboy as shown in the diagram. Water from the tap is forced in at the top, trapping and compressing the air in the carboy until the carboy is about ⅓ full.
1. **Predict** as the air is being compressed, will its temperature increase, decrease, or stay the same? **Explain** your thinking.
2. **Observe** the temperature, make records of your observations, and **Explain**.

Your teacher will then remove the stopper, allowing the air to expand.
3. **Predict**: will the air temperature increase, decrease, or stay the same? **Explain** using pictures and words.
4. Record your **observations**, and **explain** them using pictures and words.

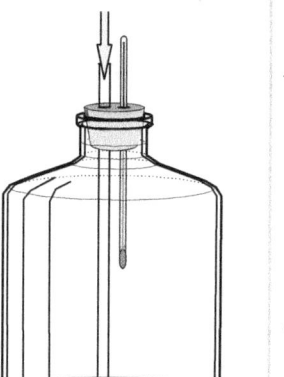

What Are We Thinking About?
- The air in this experiment is compressed to ⅔ its original volume. That is like taking an air parcel at 2.8 km, and allowing it to descended to the surface of the earth.
- When the air in this experiment is allowed to expand from ⅔ to full volume, that is like taking an air parcel from the surface up to an altitude of 2.8 km.
- Moist air does not change temperature as much as dry air.

Questions For Later...
1. You are pumping up a tire with a bicycle pump. Does the air get hotter or colder has you compress it? *Try it!*

2. An air horn contains compressed gas. Does the air horn get hotter or colder as you release the gas and allow it to expand to blow the horn? *Try it!*

© Ross Lattner Publishing   www.rosslattner.ca

Energy and the Weather

Name:
Date:

Focus Question: Write the question that you are trying to answer.

1, 2 *Predict* the temperature change when your teacher squeezes the air in the carboy. *Explain* with words and pictures.	**3,4** *Observe* the temperature change as the air is compressed into a smaller volume. *Explain* using words and pictures.
1,2 *Predict* the temperature change when your teacher allows the air in the carboy to expand. *Explain* with words and pictures.	**3,4** *Observe* the temperature change as the air expands into a larger volume. *Explain* using words and pictures.

© Ross Lattner Publishing   www.rosslattner.ca

Explaining the Weather

10 Academic Science Lab Manual

Activity 1.10: How Far Will an Air Parcel Rise?
Do you Remember? Write the 6 propositions of the Concept of an Air Parcel.

1. _____
2. _____
3. _____
4. _____
5. _____
6. _____

What's The Question? If cold air tends to sink, why doesn't the cold air above us just sink to the ground? Why doesn't the warm air near the ground just keep on rising forever? Now that we know that an air parcel cools as it rises and expands, we can more accurately predict what happens to a warm air parcel as it begins its journey skyward. *How far will an air parcel rise?*

What Are We Thinking About?
- Whenever particles move apart against an opposing force, their temperature falls. When particles are forced closer together, their temperature increases.

- If an air parcel rises 1.0 km, it expands and gets colder by 10°C. When an air parcel sinks 1.0 km, it is compressed and gets warmer by 10°C. *This cooling or warming will take place regardless of the surrounding temperatures!*

What Are We Doing? On the opposite page, you see two scenarios. In each scenario, a parcel of air is found near the ground.

1. Determine whether the parcel will rise.

2. If the parcel rises, move it 1 km, and calculate its new temperature.

3. Determine whether the parcel will rise, sink, or rest.

4. If necessary, repeat steps 2 and 3.

Questions For Later...
1. A *temperature inversion* (T inversion) is a situation in which the surrounding air increases in temperature as you ascend. Find the T inversions on the opposite page. What happens to air rising air parcels when they meet a temperature inversion? Explain.

Energy and the Weather

Name:
Date:

1 One nice spring day, an air parcel on the ground is heated to 30°C by the sun, and begins to ascend. Follow the air parcel upward as it rises, 1.0 km at a time. An air parcel cools 10 °C for each km that it rises. At what point does the parcel stop rising? Show all of your work.

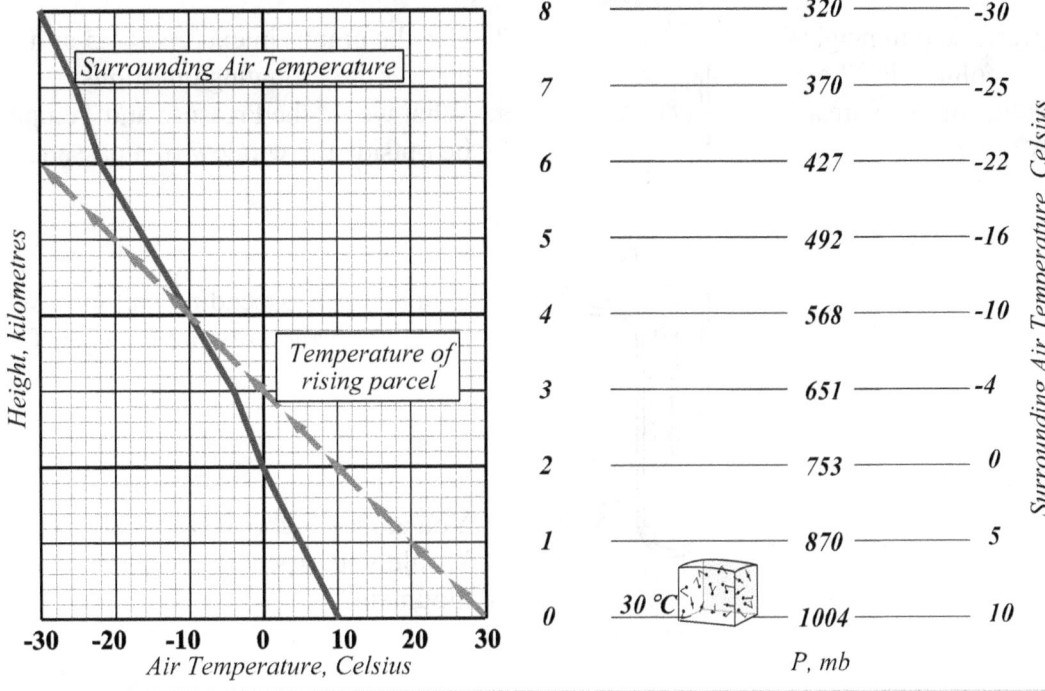

2 On a hot summer day, this air parcel is heated to 30°C on the ground. Follow the air parcel up as in the last question. How far does it rise? Explain.

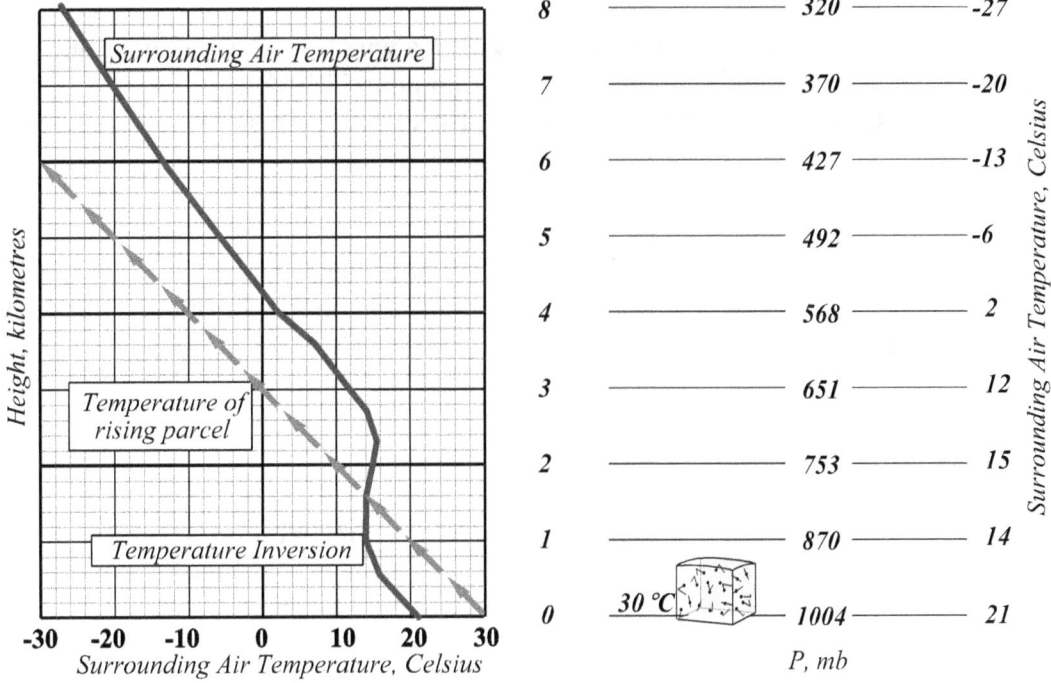

Conc

The Grade Ten Daily

All the news that's fit to print... and then some

Quiz 1.11: Energy and the Weather

1. The difference h in heights of the water column is 32 cm. What is the air pressure at point A?

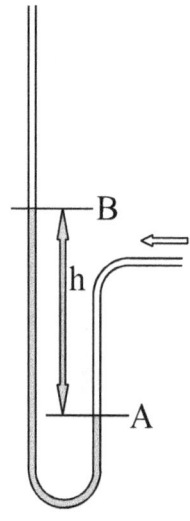

2. Use the graphs in exercises 1.5 and 1.6 to answer the following questions:
   a. What would the Pressure and Temperature to be outside the window of a passenger jet?

   b. You are in an ascending balloon. What would the altitude and pressure be when T = 0°C?

   c. You are 4 km above sea level in the Andes. Would you expect to see snow on the mountain?

Date:

Date:

3. You are inside an airplane flying at 3000 m. Outside, there are clouds, so the humidity must be 100%. Use the tables in the previous exercises to find:

   a. the temperature outside

   b. the pressure outside

   c. grams of water per 1.00 kg of air

4. It's a windless 32°C noonday July, and the humidity is 60%.
   a. How many grams of water are in each 1.00 kg parcel of air?

   b. What is the dew point? (i.e., the temperature at which the RH% becomes 100% and dew begins to form)

   c. If the temperature falls to 16°C that evening, how many grams of dew will condense out of each 1.00 kg parcel of air?

Date:

Date:

Quiz 1.11: Energy and the Weather

Name:

Cool (15°C), dry (20% RH) air parcels are drifting slowly along the ground. One parcel passes over a patch of dark brown garden on a warm spring day, and is changed to 21°C, and 40% RH.

5 What is the mass of water carried in the air parcels, both before and after heating?

Date:

6 How much energy would it take to both heat the air and vaporize enough water to change the cool parcel into the warm one?

Date:

7 The warm air parcel rises to an altitude of 2000 m, cooling due to expansion as it rises.

a. What temperature will the parcel be?

b. What mass of water, if any, will condense at that temperature to form clouds?

Date:

8 Suppose that the warm parcel of air rises to a temperature of −14°C, and 5.2 g of water condenses out to form ice crystals.

a. How much heat will be released as the water vapour condenses into ice crystals?

b. If the heat from (a) was used to heat the 1.00 kg parcel of air, what change in temperature (ΔT) could you expect?

Date:

All the news that's fit to print... and then some
The Grade Ten Daily

Quiz 1.11: Energy and the Weather   Name:

It is a hot day, 27°C. You have a can of cold 6°C cola sitting on a shady table. You notice that dew begins to form on the can.

9 What is the lowest that the humidity could be, expressed as RH%?

Date:

10 You notice that the cola gradually warms up, and the dew stops forming. The temperature of the cola is now 20°C. What was the exact RH% of the warm summer air?

Date:

11 A parcel of the air near your table ascends, cooling 1.0°C for each 100 m it rises. How high will it rise before a cloud begins to form?

Date:

12 It's so hot and muggy! You go down to the basement and turn on the air conditioner.

a. As the temperature falls, what happens to the relative humidity inside the basement?

b. Why does the old couch in the basement feel so clammy?

Date:

All the news that's fit to print... and then some

The Grade Ten Daily

Quiz 1.11: Energy and the Weather   Name:

You are camping in early May. The sky had been clear until 10:00 AM. Then the sky became steadily more cloudy, and the temperature fell, even though there was no wind. Now it has begun a slow, steady, dripping rain. Outside your tent, the T = 5°C and RH = 100%.

13 You start to play cards in the tent, and in a few minutes, the temperature in the tent reaches 15°C. What is the new RH % ?

Date:

14 The radio tells you that the atmospheric pressure has dropped. How is falling pressure related to cool, cloudy, rainy weather?

Date:

15 The rain clears off the next day, but there is a lot of water on the leaves. The temperatue climbs to 22°C, RH = 30%. Compare the relative density of the air now, and as it was during the rain.

Date:

16 A great mass of dry air has moved into your area, squeezing the air parcels all around you. What will happen to their temperature? Their relative humidity? Explain.

Date:

© Ross Lattner Publishing   www.rosslattner.ca

10 Academic Science
Lab Manual

Explaining the Weather

Lab 2.1: The Concept of an "Air Cell"
Do you Remember? List the 6 propositions of the concept of an Air Parcel.

1. _____
2. _____
3. _____
4. _____
5. _____
6. _____

What's The Question? The term "air cell" is usually used to denote a circulating mass of air. The "cell" is the entire current of air, consisting of a huge number of air parcels. *How does an air cell operate? What factors drive the circulation of air?*

NB: Answer all questions below with both particle diagrams and sentences.

What Are We Thinking About?
1. Review the exercises for the relative density of air, relative humidity, cooling due to expansion, and change of state of water.

2. Examine the diagram at right. Where inside the box would particles be moving fastest? Slowest?

3. Where in the box would heat be transferred by conduction? By radiation? By convection?

4. Where in the box does water change state? Describe the changes of state, and the energy that must be transferred.

5. Where in the box would the air particles be farthest apart?

6. Where in the box would water vapour be most plentiful? What effect would that have upon the density of the air?

7. Where would density of the air be greatest? Least? Explain.

What Are We Doing?
1. Set up a cardboard box as shown. Fill the aluminum cans ¾ crushed ice and ¼ salt. Suspend with strings or wires from the top. Place a beaker of hot water on the hot plate, and set the hot plate to low. Light an incense stick in a test tube to provide a little smoke. Seal the front with some cling wrap.

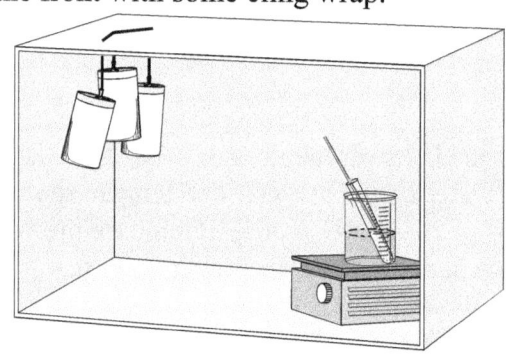

2. Study the diagram above. **Predict** the direction of air circulation. **Explain** your prediction, using particle diagrams and other ideas from this course.

3. **Observe** how the air circulates. **Explain** the circulation, using the particle theory, energy of change of state, modes of heat transfer, and other ideas you learned in this unit.

© Ross Lattner Publishing   www.rosslattner.ca

Air Circulation

Name:
Date:

Focus Question: Write the question that you are trying to answer.

1 *Predict*	2 *Explain*
3 *Observe*	4 *Explain*

Questions For Later...   *NB: you must use particle diagrams and sentences!*

1. Estimate the mass of water that evaporated. How many Joules of energy were added to the water?

2. Estimate the mass of water that condensed onto the cans. How many Joules were transferred?

3. Where, inside the box, does gravity exert the greatest force upon 1.0 L of air? Explain.

© Ross Lattner Publishing   www.rosslattner.ca

10 Academic Science Lab Manual

Explaining the Weather

Activity 2.2: Air Cells, and the Formation of Cumulus Clouds

What's The Question? In the last exercise, you developed a model of "air cell" circulation based upon that heating, cooling and changes of state. Apply that model to the formation of cumulus clouds on a summer day. *Use diagrams, calculations and sentences to answer each question.* Label the diagram below as fully as possible.

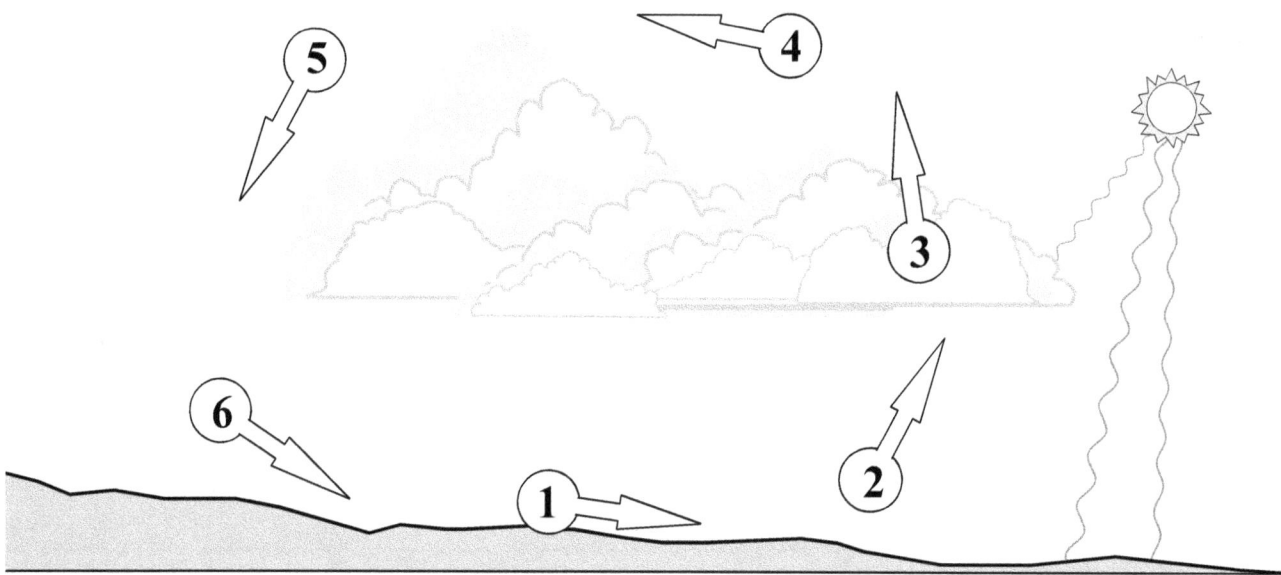

1 Cool air near the ground warms from 10°C, 20% RH to 26°C, 50% RH. Calculate: a. the mass of water *added* to each 1.00 kg parcel b. the energy used to both heat the air, and vaporize the water during the changes c. the density of the air parcels before and after.	2 As the air rises, it does not mix with surrounding air. Instead, it cools 1.0 °C for each 100 m it rises, due to expansion. a. What is the dew point for the air parcels? b. At what temperature and elevation does the rising humid air reach the dew point?

The cool, dry air falling in (5) forms a cascade of air gives cumulus clouds their smooth, rounded, sculpted shape. You can see evidence of the falling air near the edges of the clouds.

© Ross Lattner Publishing   www.rosslattner.ca

Air Circulation

Name:
Date:

3 The air parcel rises and expands, eventually cooling to −2°C at an altitude of 2.5 km.
a. What mass of water would condense out?

b. How much energy would be released by the condensing water to form cloud droplets?

4 If all of the energy of the condensing water (3) was used to raise the temperature of the air parcel, what would its new temperature be?

Of course, this cannot be the case. The air parcel must lose heat by radiation ....

5 The parcel continues to rise and mix with the surrounding cold, dry air. It radiates thermal energy in all directions until it is −2°C, and 50% RH.
a. Draw a particle diagram showing air particles radiating heat away.
b. Calculate the relative density of the air parcel.

6 The −2°C, 50% RH air parcel sinks from 2.0 km to ground level, becoming more compressed as it falls. If the air parcel warms by 1.0°C for every 100 m descent
a. What is the new temperature at the ground?

b. What is the RH% of the air at the ground?

The tiny cloud droplets fall until they pass through the bottom of the cloud into warmer air where they simply evaporate. Cumulus clouds usually disappear in this way each evening. In this way, the relative humidity of the atmosphere gradually increases over a period of several days.

10 Academic Science Lab Manual

Explaining the Weather

Activity 2.3: The Chinook

What's The Question? The famous chinook is a wind that sweeps down from the mountains across the foothills. Overnight, the chinook can change Alberta from winter to spring. *How does the chinook "work?"* Answer each question, and label the diagram below as fully as possible.

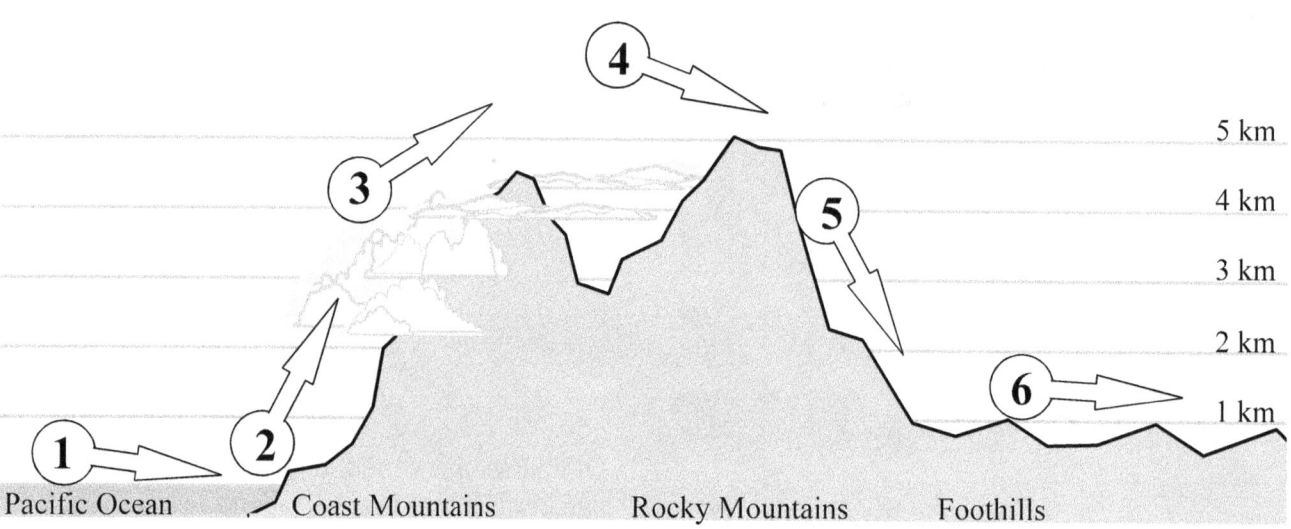

Pacific Ocean — Coast Mountains — Rocky Mountains — Foothills

1. When the Pacific air is pushed east in the spring, it is about 16°C and 100% RH.

   a. What mass of water is contained in a 1.00 kg parcel of air?

   b. What is the relative density of the air parcel?

2. As the moist air is pushed toward the mountains, it is forced to rise quickly. For each 1 km it rises, it cools 10°C.

   a. Calculate its temperature at 3 km.

The west coast of BC is home to rain forests. Why?

Air Circulation

Name:
Date:

3 Between situations (1) and (2) left, 6.0 g of water vapour precipitate out as rain, releasing the heat of condensation.
a. How much heat is released by condensation?

b. If the air parcel was –14°C before the water condensed as rain, what is its T after being warmed by the heat of condensation?

4 The air continues to rise to 5 km, and –33°C. At this temperature, another 5.0 g of water sublimes out as snow.
a. How much heat is released as the water vapour changes to solid snow?

b. Assume that the heat released by the formation of snow heats the air parcel. What is the final temperature of the parcel?

5 The air parcel, –19°C begins to sink on the eastern side of the mountains. Its T increases by 10°C for each km of descent.
a. If the parcel descends from 5 km to 1 km, what will be its temperature in the foothills?

6 The air has lost all but 0.3 g of its water in steps 3 and 4.
a. What is the relative humidity of the chinook as it rushes across the foothills?

Where is the best skiing in the Rockies?

© Ross Lattner Publishing   www.rosslattner.ca

Explaining the Weather

10 Academic Science Lab Manual

Activity 2.4: Air Masses and Atmospheric Pressure

What's The Question? What is the mass of air over a 1 km ×1 km square, 773 m high? (See bottom of page) In weather systems, an air mass covers thousands of square kilometres, and may be 10 km high. It can also move great distances over the surface of the earth. *Answer each question as thoroughly as possible in the following boxes.*

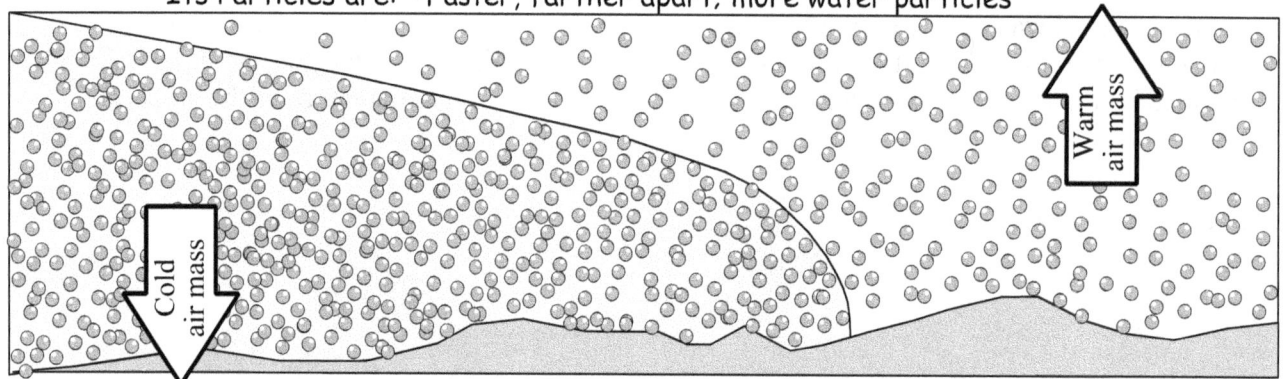

Warm air mass is: Less massive, greater volume; less dense, tends to ascend
Its Particles are: Faster, farther apart; more water particles

Cold air mass is: More massive, less volume; more dense, tends to descend
Its particles are: Slower, closer together; fewer water particles

1 Because cold air parcels are smaller and more dense, you can stack more parcels into a sky-high pile. On a cold, dry day, you might have 10 200 air parcels overhead, providing a sea level pressure of 1020 mb.	2 Because warm air parcels are expanded, you can't fit as many into a sky-high pile. In a warm, humid air mass, you might have about 9 800 parcels overhead, with a total mass of 9 800 kg, giving a sea level pressure of 980 mb.

Virtually all of the density and pressure changes occur within the troposphere. Imagine an air parcel in your house at 20°C, 50% RH and 1010 mb. Suppose each of the air masses above passed over your house. What would happen to the air parcel inside your house as the air mass passed overhead?

As the cold air mass passed overhead...	*As the warm air mass passed overhead...*
• The parcel would be compressed, and the particles would get slightly closer.	• The parcel would expand, and particles in your air parcel would get slightly farther apart.
• Temperature would increase by about 1 °C per 10 mb pressure increase.	• Temperature would fall by about 1 °C for every 10 mb loss in pressure.
• T would increase to 21°C. Relative humidity would fall to 47 %.	• T would fall to 17°C. Relative humidity would rise to 61%.
• Water particles would be slightly less likely to stick to each other, and the air parcel in your house would become clearer.	• Water particles would be slightly more likely to stick to each other and condense, making the air slightly cloudier.

The air above a 1 km × 1 km square, 773 m high, would contain 1000 × 1000 × 1000 parcels. That's 1 000 000 000 kg, or 10^9 kg. Another way of saying it: 1 000 000 tonnes, or 10^6 t. A million tonnes!

Air Circulation

Name:
Date:

3 A cold, dry air mass is sitting overhead. The air pressure outside is 1025 mb. How many 1.00 kg air parcels are overhead?

4 You are in a low pressure zone. The air pressure is 990 mb. How many 1.00 kg air parcels are sitting overhead?

5 Gravity will exert a greater force on a cold, dry air mass. Change the diagram below, to show how both of the air masses move, and what they might look like in a few hours.

Explain your thinking:

Explaining the Weather

10 Academic Science Lab Manual

Activity 2.5: Warm Fronts

What's The Question? If a mass of warm, moist air originating over oceans and lakes overtakes a body of cooler, dryer air, a warm front is created. *How does a warm front behave?* Answer each question as carefully as you can. Use diagrams and calculations as needed.

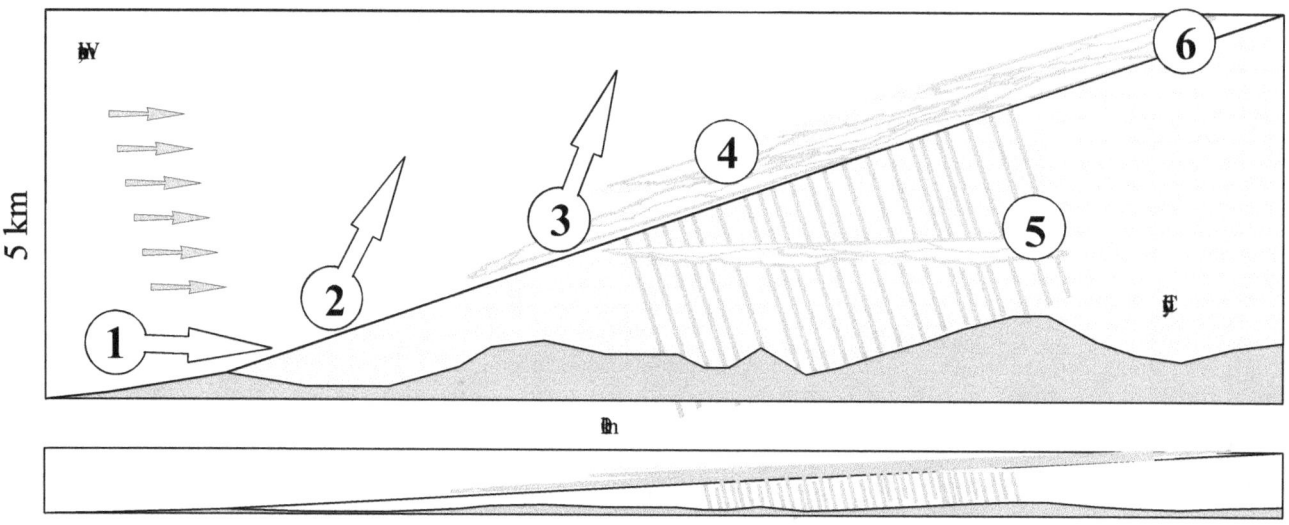

In the diagram above, the warm air mass is moving from the left, rising up and over the cooler air. It also pushes the cooler air to the right. In the upper diagram, the vertical dimension is exaggerated for clarity. The lower diagram shows the altitudes in a more realistic scale relative to the ground. Perhaps you have seen the sky look like the lower diagram.

1 An air parcel in the warm front is about 20°C, 60% RH. Just inside the cold air mass, the air is 6°C and 20% RH. a. What mass of water vapour is in each parcel? b. Calculate the densities of the two air masses. Why does the warm air rise over the cold air at the front?	**2** The moving warm is lifted up and over the cold air to higher altitudes, where it expands as it ascends. The warm air cools about 10°C for each kilometre it is lofted. a. At what altitude does cloud begin to form in the warm air?

© Ross Lattner Publishing   www.rosslattner.ca

Air Circulation

Name:
Date:

3 As the warm air parcel moves slowly to the right, it gradually rises.
a. Find the temperature of the warm air parcel at 2.0 km.

b. What is its maximum water content? What mass of water must have precipitated out? *This light rain, or drizzle, lasts for many hours.*

4 As droplets of rain fall through the cool air mass, some of the water evaporates.
a. Suppose that 1.20 grams of drizzle evaporate into each of the cool air parcels as the drizzle falls through it. How much heat would be needed to vapourize 1.20 g of water?

b. If the heat to vapourize the drizzle came from the air parcel, calculate the final temperature and humidity of the parcel.

5 As the warm air moves in overhead, the pressure falls from 1005 mb to 995 mb.

a. As the barometer falls 10 points, the air parcels expand, causing a 1.0°C decrease in temperature. What is the new temperature and humidity of the air parcels near the ground?

As the warm front passes overhead, new layers of cloud form in the lower air. The temperature falls, and fog forms, increasing the gloom.

6 At very high altitudes (e.g. 10 km) the temperature can be –40°C. Even the tiny mass of water still present freezes out into fine ice crystals. The wind at that altitude is quite fast.
The day before the rest of the warm front reaches you, which one of these two kinds of cloud would you see? Explain.

© Ross Lattner Publishing   www.rosslattner.ca

Explaining the Weather

10 Academic Science Lab Manual

Activity 2.6: Cold Fronts

What's The Question? If a mass of cold air originating at high altitudes and high latitudes overtakes a body of warmer, moister air, a cold front is created. *How does a cold front behave?* Answer each question as carefully as you can. Use diagrams and calculations as needed.

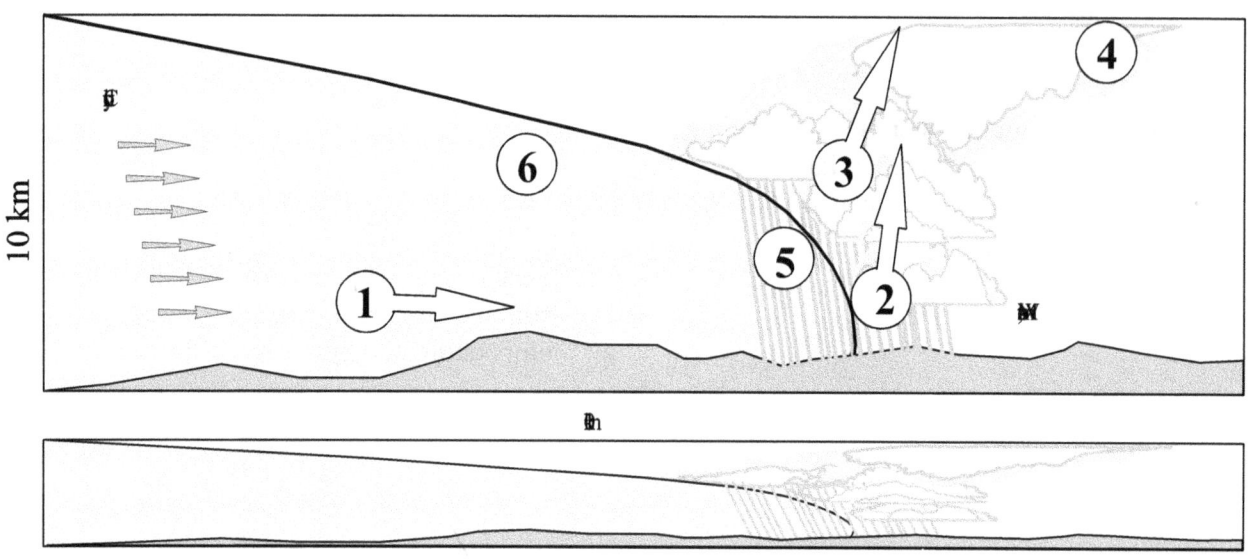

Cold air masses are pulled downward more strongly by gravity. They tend to move more quickly as they spread out, creating a steep front edge. The warm, moist air is quickly forced upward to high altitudes. The greater the density difference, the more extreme are these trends. The vertical dimension is exaggerated in the upper diagram; the lower diagram is a more realistic scale relative to the ground.

1 On a muggy day in August, the stationary warm air mass is 30°C, 70%RH. The moving cold air front is 12°C, 20% RH, and clear. a. What mass of water vapour is in each parcel? b. Calculate the densities of the two air masses. c. Why does the cold front pass under the warm air mass?	**2** The warm air is pushed to higher altitudes. A warm air parcel is lifted to 3 km. a. What would its new temperature be? b. Suppose that 8.0 g of water condense out as cloud droplets. How many Joules of heat energy would be released into the parcel? *(Most of this heat is radiated away)*

Air Circulation

Name:
Date:

3 The air parcel in (2) has lost enough heat to reach 20°C at 3 km. Because it is warmer than its surroundings, it rises to 8 km.

a. The parcel cools by expansion as it rises. What is its temperature at 8 km?

b. Assume that all remaining water vapour is frozen out as ice crystals (11. g). How much energy would be released?

Most of this energy is radiated away, but some creates powerful vertical winds.

4 The air parcel may be lofted even further, carrying fine ice crystals. This is the appearance of an anvil cloud, pushing up into the stratosphere.

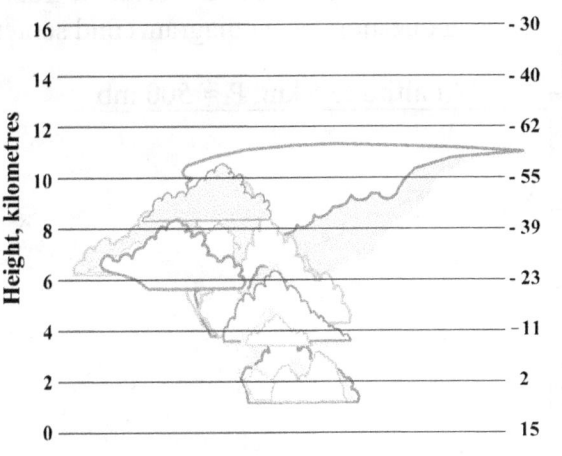

Why doesn't the air parcel rise to 16 km?

5 A typical thunder cloud is about 2 km wide × 10 km long × 10 km high, and contains about 1×10^{11} kg of air. Recall that the original air parcels were 30°C and 70% RH.

a. What is the total mass of water in a thunder cloud that could condense as rain?

The rainfall from a thunder head is very brief and heavy, easily out-soaking Niagara Falls.

6 Shortly after the cold front passes, the air pressure increases, the cool air near the ground becomes warmer, and the sky becomes nearly clear. Explain why this happens.

Now the process of cumulus cloud formation can begin all over!

Explaining the Weather

10 Academic Science Lab Manual

Activity 2.7: Atmospheric Pressure and Isobars

What's The Question? Suppose a rapidly moving cold front overtakes a slow-moving warm front. The warm, moist air, already moving up and over the cool air, is pushed up from beneath by the cold front. Such an event produces an *occluded* front. *How does an occluded front affect pressure?* Answer each question, using diagrams and sentences. The diagram depicts an altitude of 5 km.

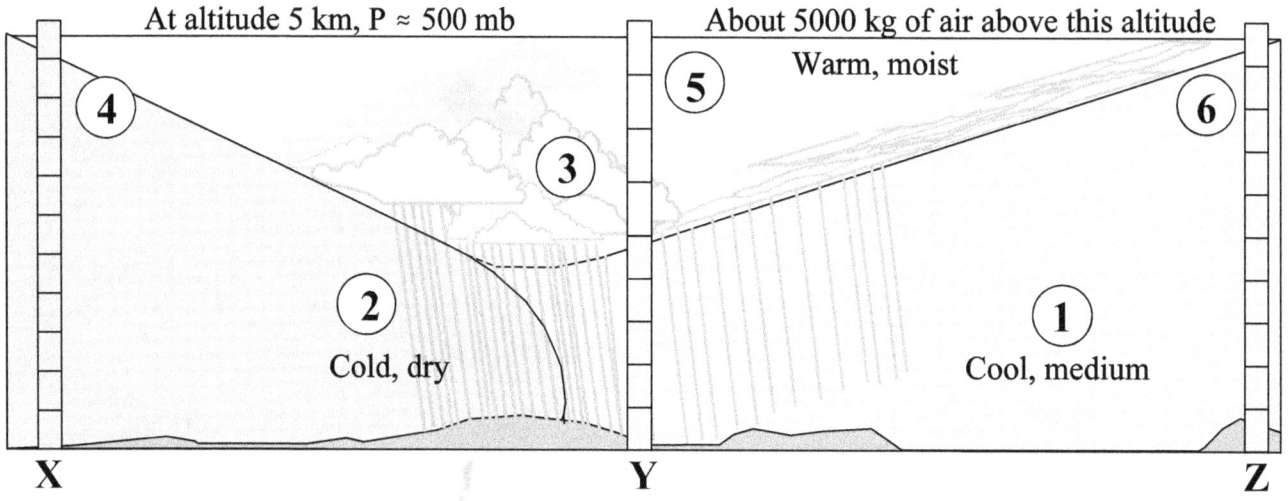

The three vertical columns (X, Y, and Z) represents three stacks of super-parcels. Each super-parcel contains 500 regular parcels and contains 500 kg of air. Colder parcels are smaller and denser. Warmer parcels are larger and less dense.

1 The cool air mass (13°C and 40% RH) is moving most slowly. Calculate the density of the air in a typical parcel.	2 The cold air mass (5°C and 20% RH) is moving quickly, driven by the downward force of gravity. Calculate the density of a typical parcel of this air.
3 What was the density of one 1.00 kg parcel of the warm air mass? (originally 26°C and 80% RH)	4 The cold front, moving at 10 km/h, pushed right under the warm front, which was moving at 5 km/h. Explain why the warm, moist air was lifted right up and away from the earth.

© Ross Lattner Publishing   www.rosslattner.ca

Air Circulation

Name:
Date:

5 Consider the three stacks of "super-parcels" marked X, Y, and Z. Each super-parcel contains 500 regular parcels, and has mass of 500 kg,

a. Which stack (X, Y, or Z) will be the most massive? Explain your reasoning.

b. Which stack (X, Y, or Z) is descending in the picture? Which stack is ascending? Explain

c. Which stack (X, Y, or Z) will exert the greatest pressure on the ground? Explain.

6 Recall that atmospheric pressure is the *force of gravity on all of the air parcels above you.*

a. Count all of the parcels in stack X. Add up the total mass of stack X. Now add the 5000 kg of air in the upper atmosphere. What is the total pressure under stack X?

b. Find the total pressure at the bottom of stack Y and stack Z in the same way.

7 The weather map below shows a cold front ▲▲▲ Being pushed eastward by a cold dry air mass in the west. At the same time, warm air is being forced northward by a large mass of semitropical air from the Gulf of Mexico, creating a warm front (⌒⌒). Where the two air masses meet, an occluded front has been created ▲⌒▲⌒ The curved grey lines indicated lines of equal pressure, or isobars.

a. Draw a reasonable curve for the 975 mb isobar.

b. Write a reasonable pressure for any unlabelled isobars.

c. Cut out a photocopy of the occluded front diagram, and fasten it with tape or glue to the dark line on the map below. The stacks should line up at points X, Y and Z.

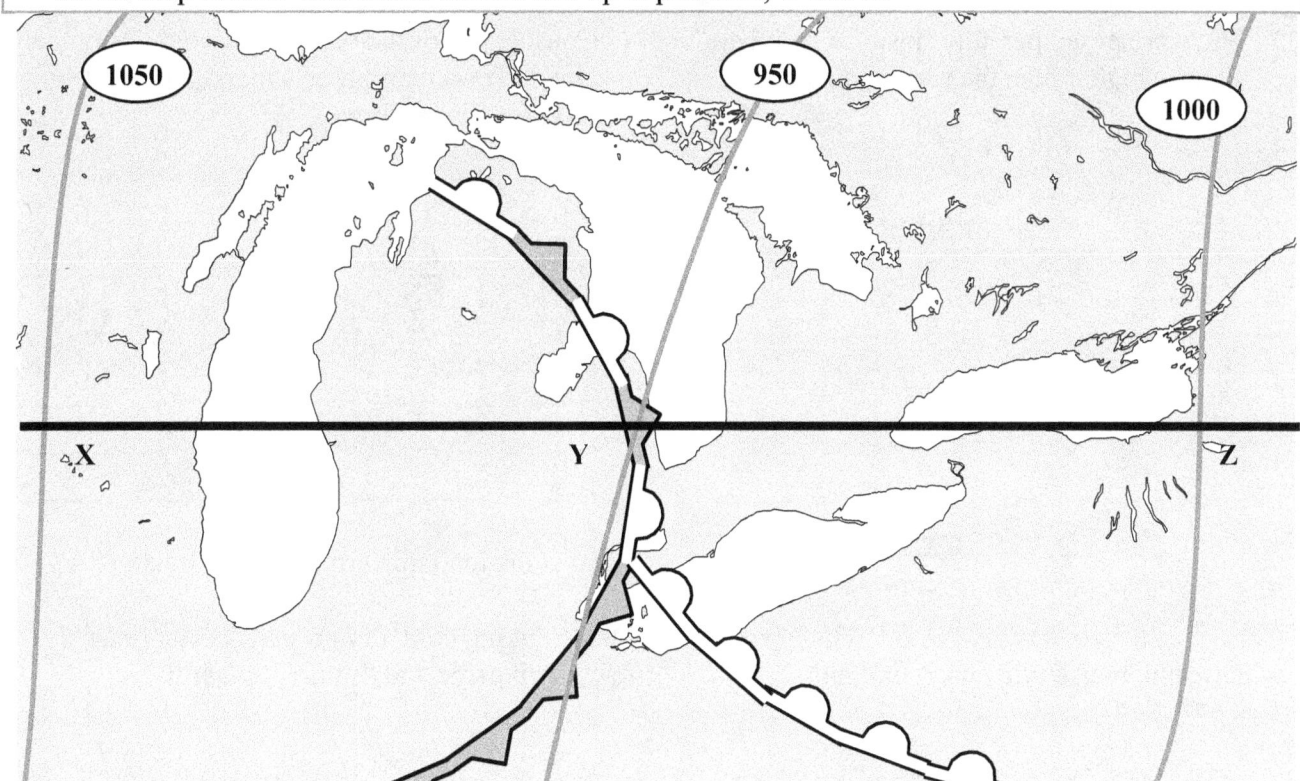

© Ross Lattner Publishing   97   www.rosslattner.ca

The Grade Ten Daily

All the news that's fit to print... and then some

Quiz 2.8: Air Circulation

1 Add arrows to both of the diagrams below to indicate the direction of air flow in the cells.

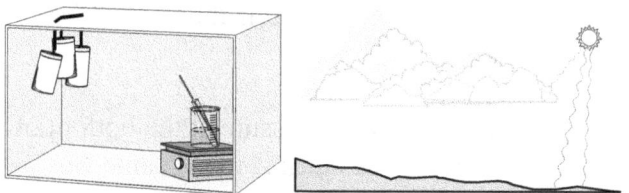

List three similarities and three differences in how these situations cause air to circulate.

Date:

2 *"For every 1. kg parcel of air that rises one metre in our atmosphere, there must be another 1 kg parcel of air that sinks one metre somewhere else."*

Is this statement true or false? Explain your thinking.

Date:

3 The warmer temperatures inside a cloud are slightly higher than the surrounding air

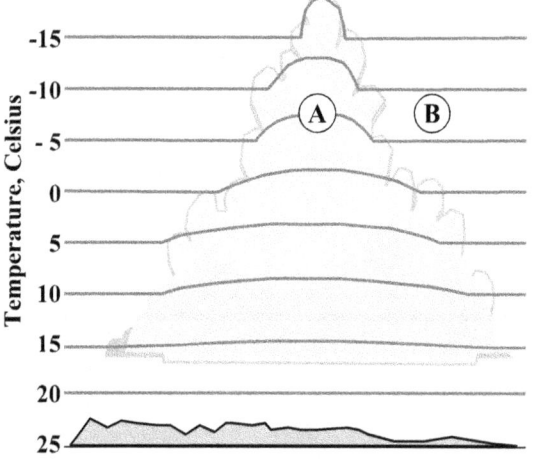

Consider the parcels at A and B. Which parcel will ascend, which will sink? Explain.

Date:

4 Consider all of the blue sky around a cumulus cloud. If each air parcels marked B has 1 unit of gravitational force:

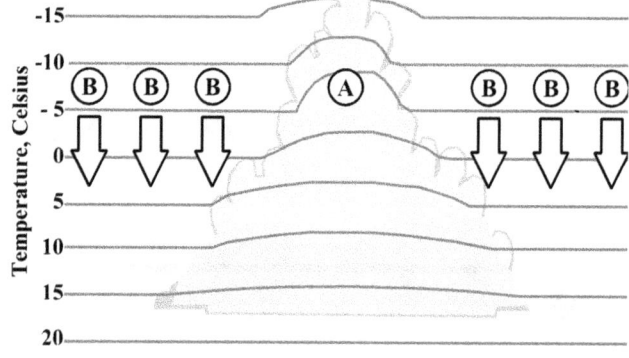

What is the total upward force on air parcel A?

Many sinking cold, dry air parcels can cause a few warm parcels to "squirt" upward!

Date:

All the news that's fit to print... and then some

The Grade Ten Daily

Quiz 2.8: Air Circulation　　　　　　　　　　　Name:

Rising and sinking air, driven by the energy of the sun, cause circulating masses of air called "cells." The cold air high in the atmosphere in the cold climates should sink, causing the warm moist air near the ground to ascend in the warmer climates. *Can there be really huge cells that cover the entire earth?* Let's look at some models to find out.

5 In this model, air heated at the equator is displaced by cold air sinking near the poles.

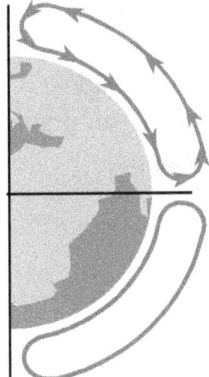

Add arrows to show the direction of circulation of both of the cells.

This model requires that cold, dense air remain aloft as it travels 9 000 km from equator to poles. How likely is that? Explain.

Date:

6 Perhaps the huge cell in (5) divides in 2. In that case, cold air sinks about half way to the poles, causing a massive downdraft at 45°.

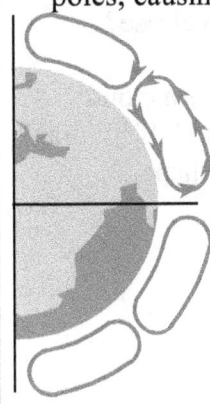

Draw arrows on the remaining cells. Note the behaviour at the poles. What is wrong with this model? Explain.

Date:

7 Since (6) is unlikely, perhaps there are three cells between equator and poles. Cold air aloft sinks at 30°, and warm air rises at 60°.

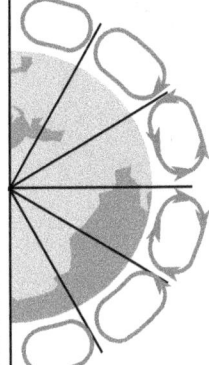

Draw arrows on the remaining cells. Is this model consistent with your understanding of circulation cells? Explain.

Date:

8 The cells in (7) are called Hadley cells, after the person who first described them.

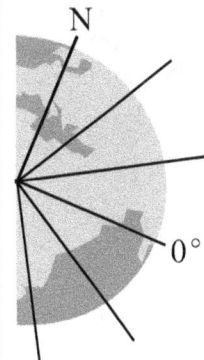

The earth is tilted at 23°, so the June sun does not strike directly on the equator. Draw rays from the sun in June.

Where will the Hadley cells be located during June?

Date:

© Ross Lattner Publishing　　www.rosslattner.ca

The Grade Ten Daily

All the news that's fit to print... and then some

Quiz 2.8: Air Circulation   Name:

In questions 9 and 10, an air parcel ascends one kilometre at a time. At each kilometre, calculate: a) the new temperature after rising and expanding; b) grams of water condensing, if any, and c) new temperature after heat of condensation until the parcel stops rising.

9 Consider a hot, dry (32°C, 10% RH) parcel of air near the floor of a gravel pit.

Height, kilometres	Temperature, Celsius
8	-27
7	-20
6	-13
5	-6
4	1
3	8
2	15
1	14
0	21

How high does the parcel rise?

Does a cloud form? If so, draw its shape.

Date:

10 Consider a hot, moist (32°C, 62% RH) parcel of air just over a wet golf course.

Height, kilometres	Temperature, Celsius
8	-27
7	-20
6	-13
5	-6
4	1
3	8
2	15
1	14
0	21

How high does the parcel rise?

Does a cloud form? If so, draw its shape.

Date:

11 A warm front is moving like a breeze over sea of cool air, and causes waves to form.

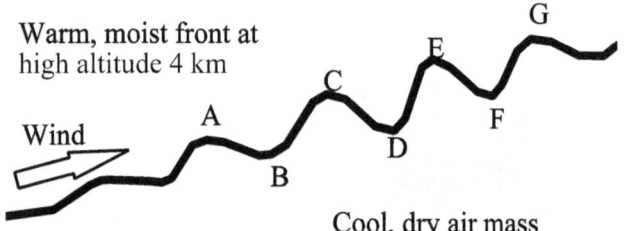

Warm, moist front at high altitude 4 km

Wind

Cool, dry air mass

At which points A - G would clouds be likely to form, as the warm moist air rose and fell over the waves? At which points would the air become clear again? Explain. (Hint: consider the chinook)

Date:

12 It's been clear since Monday. Cumulus clouds form during the day, and disappear at night. Now, at 2:00 Thursday afternoon, you can see a lot of clouds like these high overhead..

What is happening? What will the weather be like tomorrow? Explain.

Date:

All the news that's fit to print... and then some
The Grade Ten Daily

Quiz 2.8: Air Circulation Name:

You have seen so many kinds of clouds overhead. How do they form? *What can you tell about the weather by the shape and altitude of the clouds you observe?*

13 It's been wonderful weather for the past few days: clear, sunny blue sky, puffy clouds, gentle winds in the afternoon, cool and still at night. This morning, however, at 10:00 AM, you see clouds like this forming overhead.

What is happening up there? What is the weather going to be like tomorrow? Explain.

Date:

14 It has been really nice weather for February: sunshine caused melting for the past four days, nice little puffy clouds in the afternoons. The breeze picked up out of the west this evening, though, and a wall of clouds is rising over the south west horizon.

What's happening to the west? What kind of weather are you likely to get tomorrow? Explain.

Date:

15 It's been raining almost every day since October 10. Just when you think it's going to clear, more rain! The leaves have been knocked off the trees. Now, on Oct 25, it's gloriously clear. The sky is deep, deep blue; the frosty mornings are warmed rapidly to shirt sleeve temperatures. Indian summer!

What kind of air mass has moved in? Explain.

Date:

16 It's been hot for two weeks, no relief. The air is muggy, with scarcely a breeze at night. Water dribbles steadily from the air conditioners. Around sunset today, some bright clouds seem to tower way above the grey haze. Under their flat white tops, these clouds are massive, covering the western horizon from north to south.

What kind of air mass is moving in? What might happen next? Explain.

Date:

© Ross Lattner Publishing www.rosslattner.ca

Explaining the Weather...

10 Academic Science Lab Manual

Activity 3.1: We Live on a Spinning Ball

What's The Question? The earth is spinning. We are held to its rotating surface by gravity, so we do not notice its huge velocity. At the equator, the ground is racing eastward at 1670 km/h. At lunch on the 43° latitude you, your table, and your lunch smoothly speed along at 1220 km/h!

How does the spinning earth affect the movement of air?

What Are We Thinking About?

- Consider a mass of cool, dry air, sinking and spreading out in the high pressure zone in the map below. (black circles) The air parcels at the edge of the high pressure zone are being pushed outward at the gentle speed of 5 km/h. Air parcel A is moving eastward with the earth at 1200 km/h, and B is moving at 1220 km/h.

- Twenty hours later, A and B are over Georgian Bay and to Ohio, respectively.

- Nothing has caused the air parcels to speed up or to slow down in the East West direction. Parcel A is still traveling East at 1200 km/h, while the waters of Georgian Bay are only moving at 1180 km/h. The air is moving eastward *20 km/h faster* than the ground!

- Parcel B is still moving at 1220 km/h, while the shores of norther Ohio are moving at 1240 km/h. The air is moving *20 km/h slower* than the ground. This creates a wind 20 km/h toward the west.

- The high pressure system has begun to spin in a clockwise direction! Note that the spinning velocity is greater than the 5 km/h speed of the expanding high pressure zone.

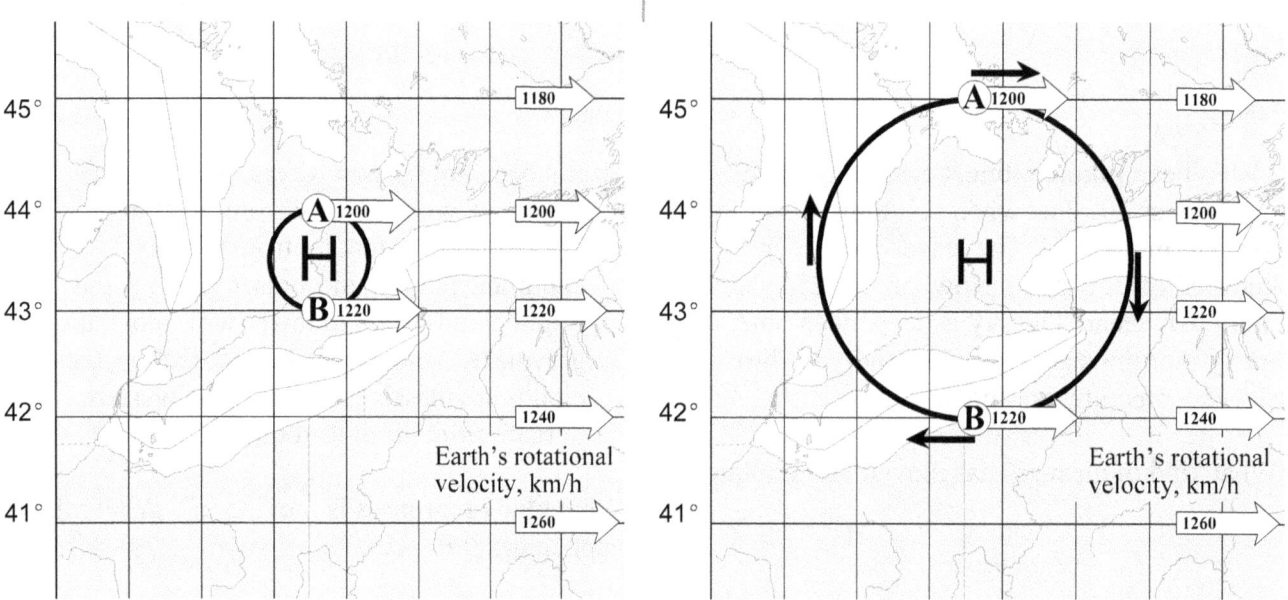

Initial high pressure system   Twenty hours later

Note that no force actually caused the air to speed up or slow down. Air parcel A drifted north with the pressure. Parcel A simply kept moving east at the same speed it had earlier. One degree north, however, the ground on the spinning ball of the earth is not moving as fast.

... on a Spinning Planet

Name:
Date:

Focus Question: Each of the air masses in the problems below will spin as they flow, due to the rotation of the earth. *Predict the speed and direction of the spinning of the air masses.* Note: the speeds you calculate will be greater than observed, because other factors are also at work.

1

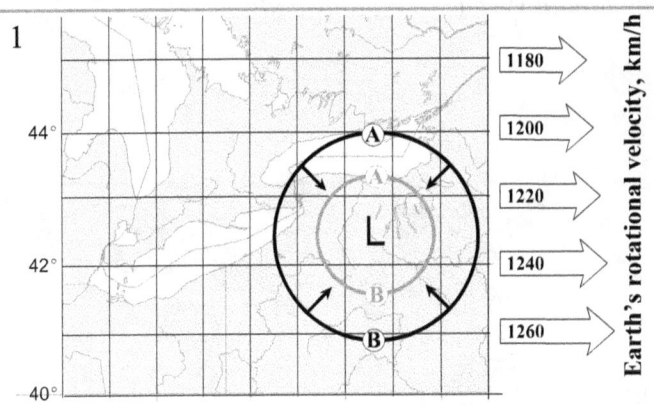

Air is forced inward into the low pressure zone from black circle to grey. In which direction will the low pressure zone spin? How fast?

2

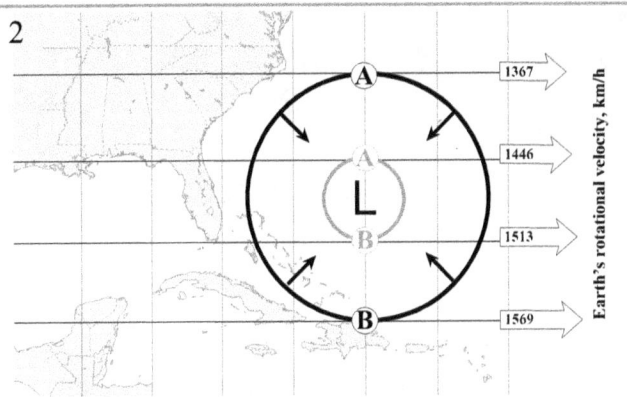

It is September. A mass of hot, humid air is forced rapidly upward over the Atlantic. Air parcels A and B (black circles) move toward the low pressure zone. A day later, they are much closer to the center (grey circles). How fast would they be traveling relative to the earth? In what direction would the low pressure zone rotate? What is this weather system called?

3

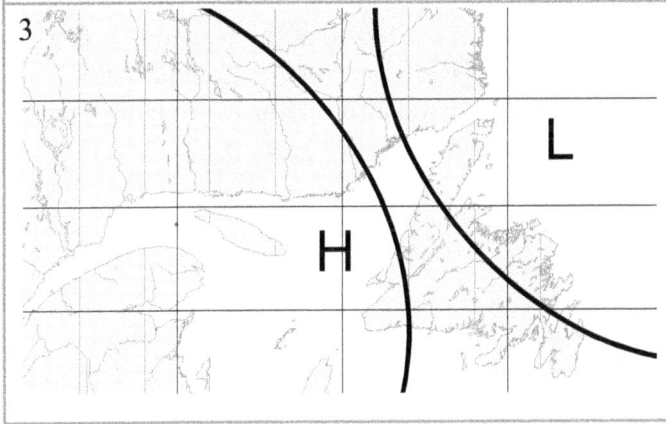

A high and low pressure zone develop adjacent to each other over Newfoundland. In what direction will the two bodies rotate, due to the spinning of the earth? In what direction will the wind blow in that area?

Questions For Later...
1. You would think that wind would blow directly from H to L pressure. What happens instead?

2. The four problems in this exercise are examples of the "Coriolis effect". What is the Coriolis effect? Your text may use the words "Coriolis force", although there is no real force.

© Ross Lattner Publishing   www.rosslattner.ca

Activity 3.2: Hadley Cells and Prevailing Winds

What's The Question? In the last exercise, you investigated how high and low pressure systems begin to rotate, due to the spinning motion of the earth. Hadley cells develop very large high and low pressure systems. *How do surface winds relate to the global system of Hadley cells?*

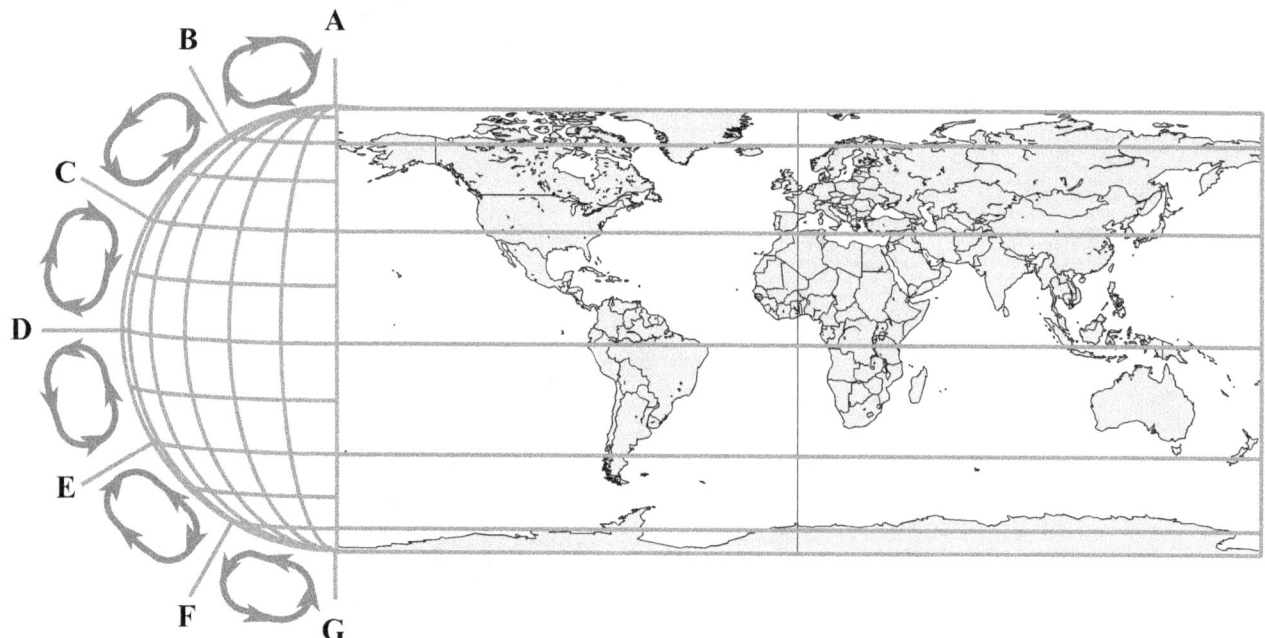

1. Descending cold, dry air causes high pressure conditions on the surface of the earth, while ascending warm, moist air causes low pressure conditions.

   Which of the letters A - G above would be regions of high pressure, and which ones regions of low pressure? Explain your choices.

2. The letter D on the hemisphere above left represents the equator. Draw a band across the map of the world at the equator. Is D a region of **High** pressure or **Low** pressure? Label the band either **H** or **L**. Explain your choice.

   Now draw bands at latitudes B, C, E, and F. Label them **H** or **L** as the case may be. Explain your choices.

The bands of high and low pressure that you have described are surely present, although they are seldom exist in the perfectly ideal form that you show on your map.

... on a Spinning Planet

Name:
Date:

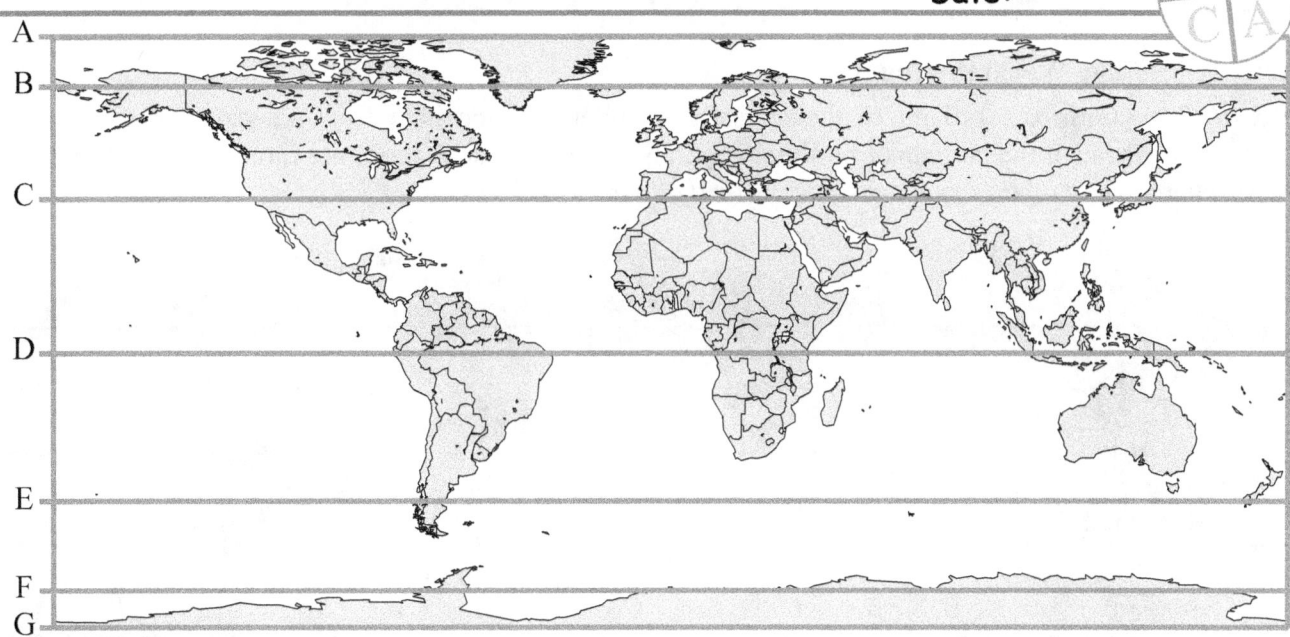

3. Suppose the earth did not spin. The Hadley cells would cause air parcels near the surface to drift either due north or due south. In which direction would air parcels between lines C and D be drifting? In the map above, draw 5 arrows to indicate surface winds due to Hadley cell circulation. Draw 5 arrows between D and E to indicate surface winds due to Hadley cells *without* the Coriolis effect. Repeat for the other four regions.

4. Now, spin the earth. Lab 3.1 showed that poleward - moving air keeps its eastward velocity. The Coriolis effect on poleward moving air creates an eastward wind. As air parcels drift toward the equator, they move more slowly than the earth beneath them, thus creating a westward wind. In the map below, draw arrows in each of the regions A-B, B-C etc to indicate the Eastward and Westward winds in those regions.

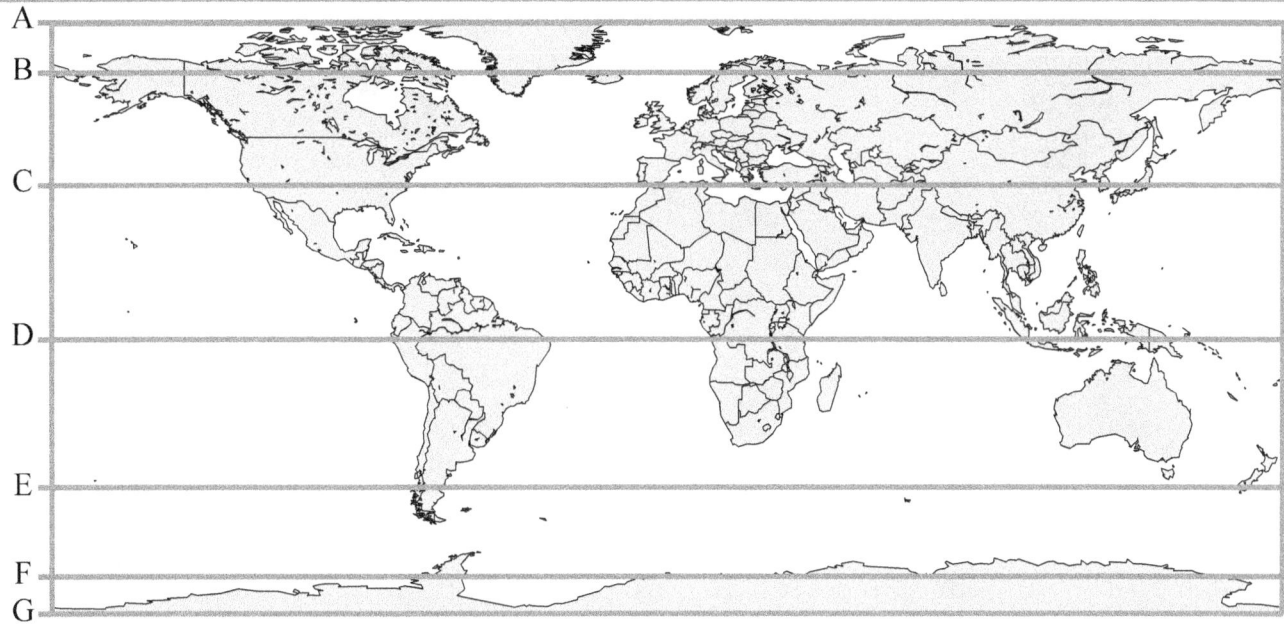

10 Academic Science
Lab Manual

Explaining the Weather...

Activity 3.3: The World in January

What's The Question? In January, the norther hemisphere is tilted away from the sun. The sun is directly overhead at the latitude of northern Australia. High pressure zones are rare in the southern hemisphere. *How do these patterns affect world climate?*

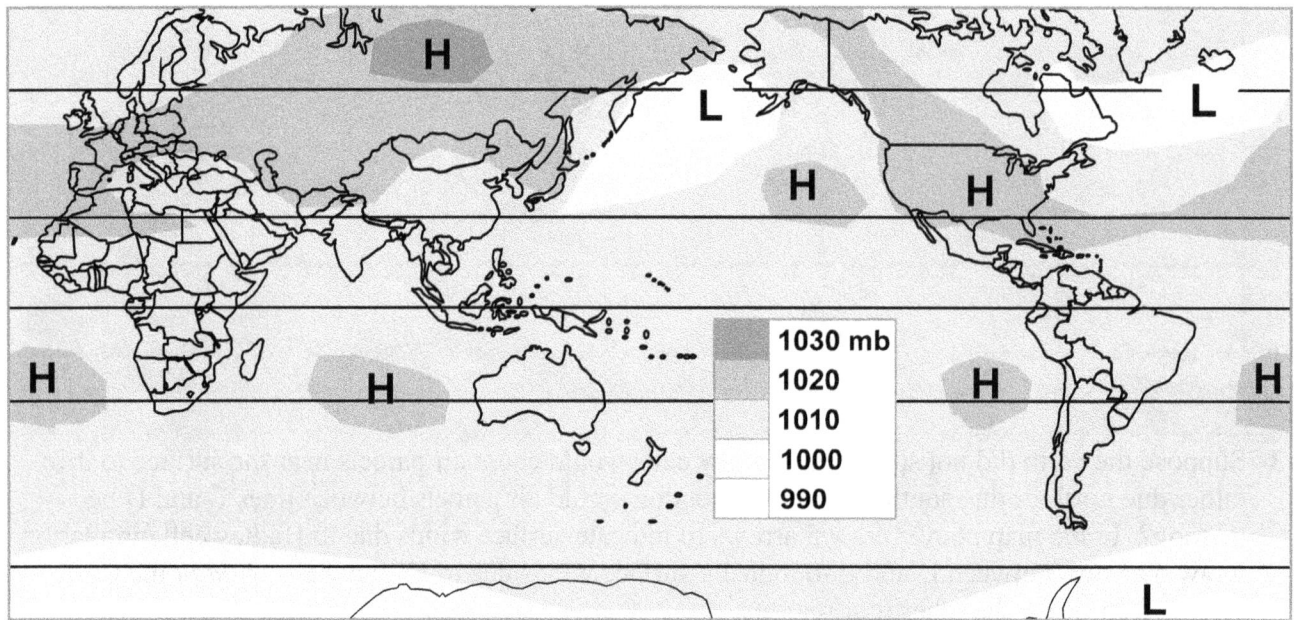

1. According to the Hadley cell theory, you would expect a downdraft of cold air from high altitudes at 30° N and S of the equator. These should appear as zones of high pressure.

   The map shows that the expected zones of high pressure are indeed present, but they are much more prevalent in the north. Why is this?

2. Draw arrows around each pressure zone to indicate the direction in which the zone rotates due to the spinning of the earth.

   a. From which direction do the winds tend to blow across most of Canada during January?

   b. That being the case, from which part of the globe does our air come in January?

© Ross Lattner Publishing

... on a Spinning Planet

Name:
Date:

3 Where is the highest pressure zone in the world in the month of January?

a. What is the air pressure over the northern Pacific Ocean?

b. Predict the speed and direction of the winds in northern Japan and the west coast of Russia at this time. Explain your prediction.

4 The spinning effect of high and low pressure zones is greatest near the poles and least near the equator. Why would that be the case? (See Lab 3.1)

5 Some references suggest that the Hadley cell system shifts southward during January. Others believe that the Hadley system remains pretty much in place, but its intensity changes.

Examine the map. Do you think the Hadley cells change position, or change relative intensity? Give reasons for your thinking.

6 The pressure patterns provided in the map are averages over many days, and years, during the month of January.

On average, do you see air ascending or descending over the continents? Over the ocean? Is there a difference between northern and southern hemispheres? What is going on?

© Ross Lattner Publishing   www.rosslattner.ca

10 Academic Science Lab Manual
Explaining the Weather...

Activity 3.4: The World in July

What's The Question? July brings summer weather to the northern hemisphere, and winter to the south. *How do these patterns affect world climate?*

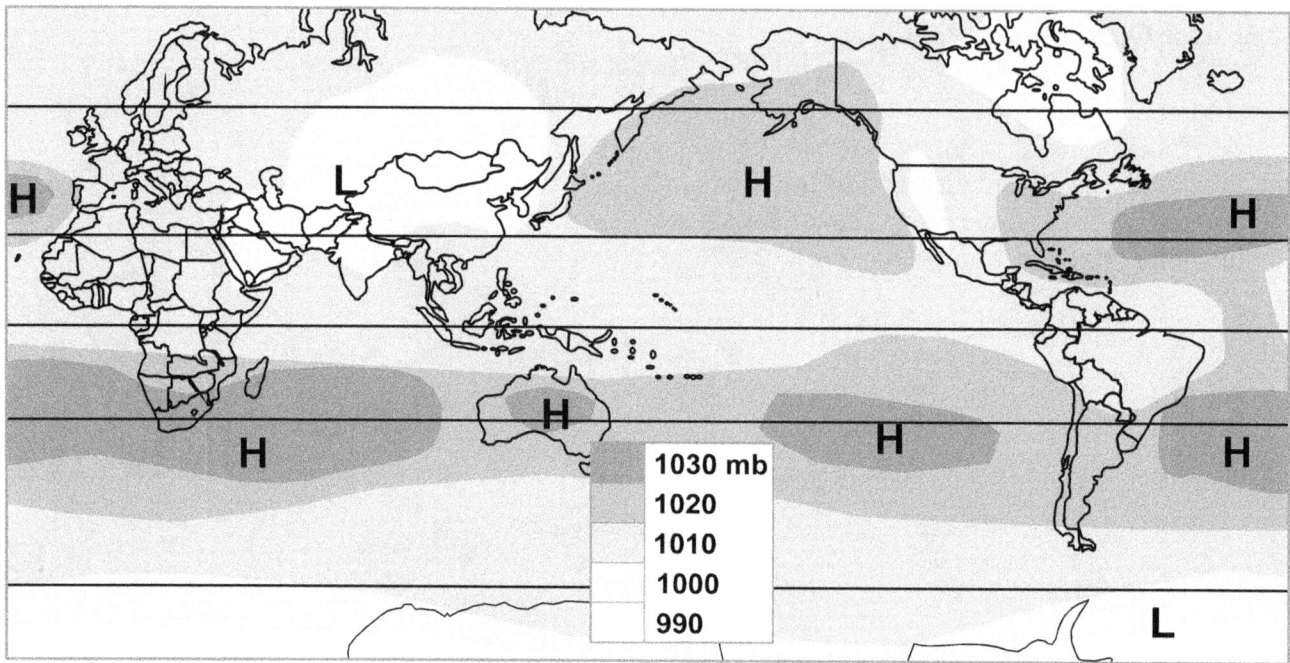

1. According to the Hadley cell theory, you would expect a downdraft of cold air (high pressure) at 30° N and S of the equator. The map shows that the expected zones of high pressure are more prevalent in the south. Why is this?

2. Draw arrows around each pressure zone to indicate the direction in which the zone circulates due to the spinning of the earth.

   a. From which direction do the winds tend to blow across most of Canada during July?

   b. That being the case, from which part of the globe does our air come from in July?

© Ross Lattner Publishing   www.rosslattner.ca

... on a Spinning Planet

Name:
Date:

3 In July, there are high pressure zones over the Pacific Ocean at 30° N. High pressure zones arise when cold air descends from higher altitudes.

a. Where does this descending cold air come from?

b. How did this pressure system influence trade routes in the days of sailing ships?

4 Spanish and Portuguese fishing trawlers claim to have the right to fish off the Grand Banks of Newfoundland. They claim to have fished there for centuries.

Does the existence of the strong high pressure zone over the middle Atlantic Ocean support their argument? Explain.

5 The North and South poles are regions of low pressure year round. The only other region of extreme low pressure is just north of India, over the Himalayas.

What would happen to hot tropical air 35°C and 100% RH, as it was forced over the Himalayas? Explain how this contributes to low pressures.

6 July is winter in the Southern hemisphere. What would the winds be like for ships sailing around the southern tips of Africa and South America during July? Explain your reasoning.

10 Academic Science Lab Manual

Explaining the Weather...

Activity 3.5: Pressure Systems and Weather Maps

What's The Question? The word "isobar" means "equal pressure." In weather maps, the isobar is a line at which the pressure is constant. In fact, an isobar almost always makes a closed loop, rather than an open line. *How do meteorologists make use of isobars on weather maps?*

What are we thinking about? A map of the Great Lakes region is shown. The atmospheric pressure is mapped as isobaric curves. All pressures are quoted in millibars. Recall that 1013 mb is the standard average atmospheric pressure at sea level. Pressure decreases with altitude. For example, the highlands of Algonquin Park are 200 m above L. Ontario. To eliminate altitude effects, all pressures have been adjusted as if all the earth was at sea level.

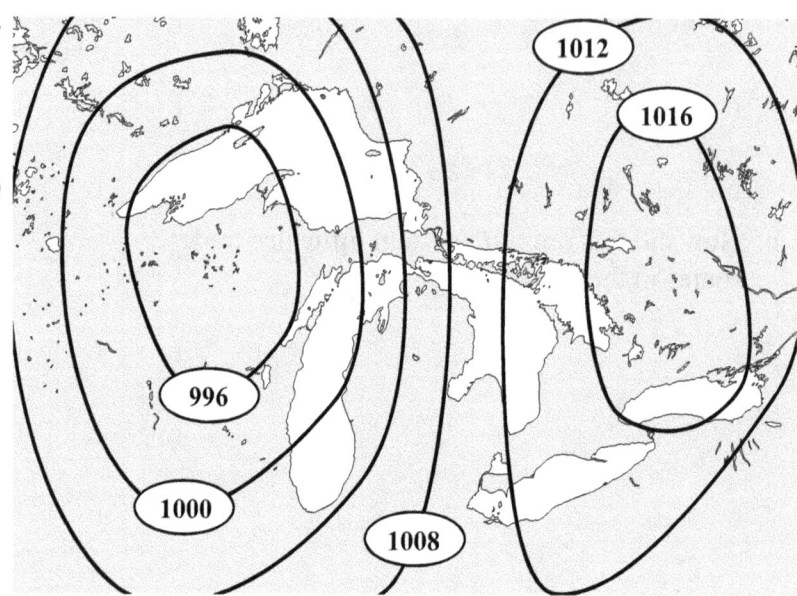

The highest pressure zone in this map, 1016 mb, extends from L. Ontario north to Kirkland Lake. The next highest pressure zone, 1012 mb, is much larger. Put your pencil anywhere in the space between the 1012 curve and the 1016 curve.

- If you stay within the space bounded by the two curves, the pressure is between 1012 and 1016 mb.

- If you stay on the 1012 boundary, the pressure is the same everywhere on the line.

- If your pencil crosses the 1012 isobar, it has moved into a zone between 1012 and 1008 mb pressure.

Remember that low pressure is related to lower atmospheric density. Low pressure air is likely to be warm, moist ascending air. High pressure air is likely to be cool, dry and descending.

What Are We Doing? For each box on the opposite page

1. **Predict** an answer, based upon your best reasoning.

2. **Explain** your answer, using diagrams, calculations and sentences to build up your argument

3. **Check** your answer with the teacher. Is your argument a strong one? Does it need some correction?

4. **Explain** your new improved argument. Correct your diagrams, calculations and sentences where needed.

... on a Spinning Planet

Name:
Date:

1. The map right shows a system similar to that on the previous page.. Mark the **High** and **Low** pressure on the map. Draw 4 arrows on each isobar line to indicate wind direction and speed. Where is the cold air mass? Where is the warm air mass?

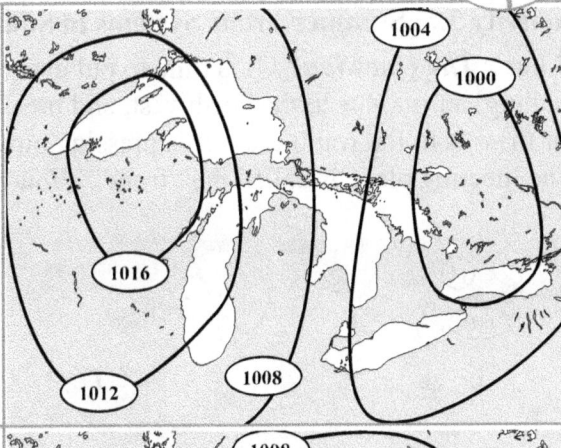

2. Is pressure over L. Superior **High** or **Low**? Is the temperature over L. Superior **Warm** or **Cold**? Mark the maps. Repeat for L Huron. Add 5 arrows to each isobaric line, to indicate wind direction and speed.

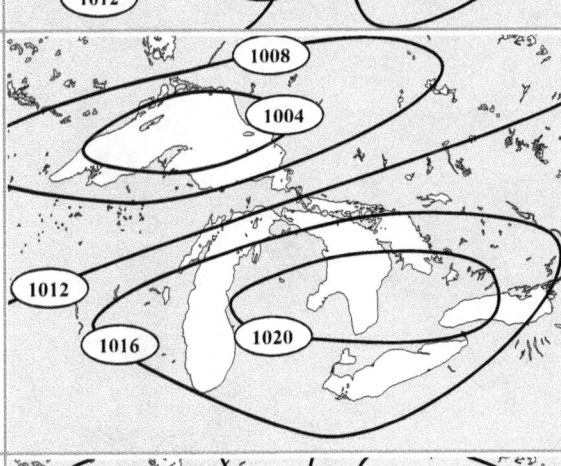

3. The air over Ottawa is clear, cold, and the pressure on the ground is high (1020 mb). Mark the pressure of the rest of the isobars, if they are at 4 mb intervals. Indicate wind direction.

   What is the weather likely to be in Chicago?

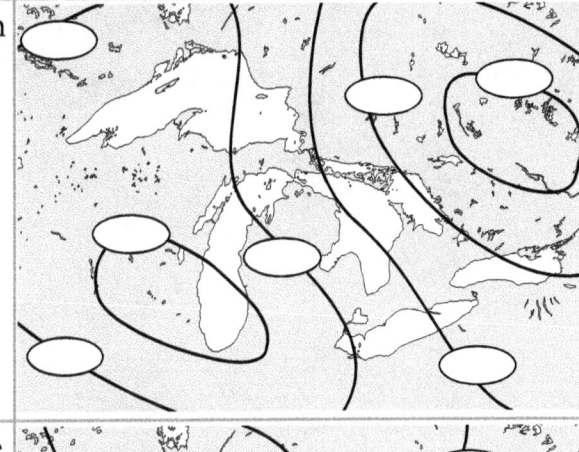

4. It's hot, 30°C and humid in Toronto, and the pressure is 997 mb. Mark the rest of the isobars at 4 mb intervals. Indicate wind direction.

   a. Predict the wind direction in Toronto.

   b. From what direction is Toronto's weather coming?

© Ross Lattner Publishing   111   www.rosslattner.ca

Explaining the Weather...

10 Academic Science Lab Manual

Activity 3.6: Summer Front Systems in Southern Ontario

What's The Question? It is time to put everything together. We will look at two air masses. A cold, sinking, air mass is to the northwest, and pressing outward due to the force of gravity. A warm, moist air mass is to the south east. It is pressing outward due to solar heating and expansion from the south. The meeting place is called the "front". *How and why does the front evolve over time?*

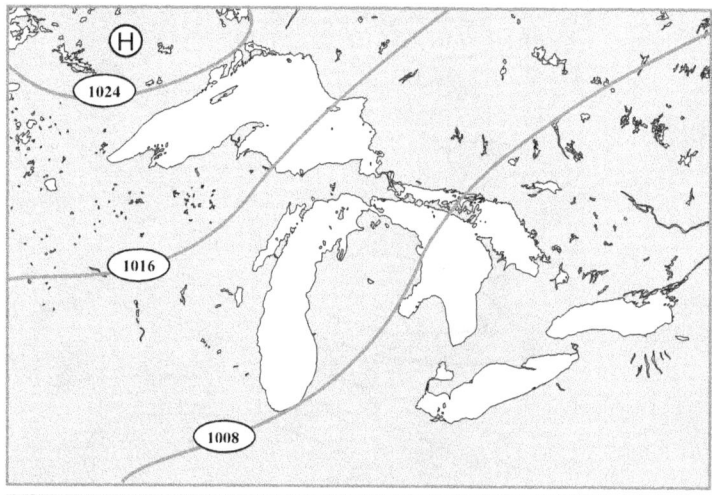

1. A cold air mass north of L. Superior is expanding outward, driven by the force of gravity. A warm air mass south of L. Ontario is expanding northward, driven by other cold air masses not on this map.

In which direction would these air masses spin? What would be the direction of the wind along the isobaric pressure lines?

This situation is unlikely to persist: Somewhere, air is going to start to move.

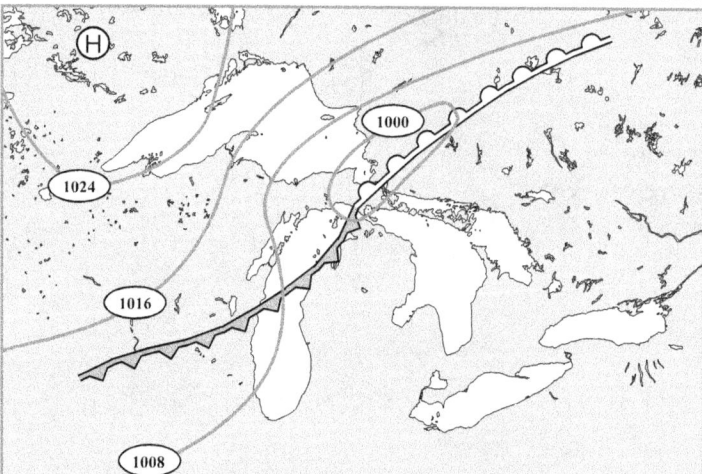

2. Some cold air moves under the warm air mass, creating a cold front (marked with the ▲). At the same time, warm air is forced northward, creating a warm front (⌒).

Clouds begin to condense where the fronts are moving most quickly. Heat from the condensation is warming the air in that regions, creation a zone of low pressure. What is the direction of the wind around the developing low pressure zone?

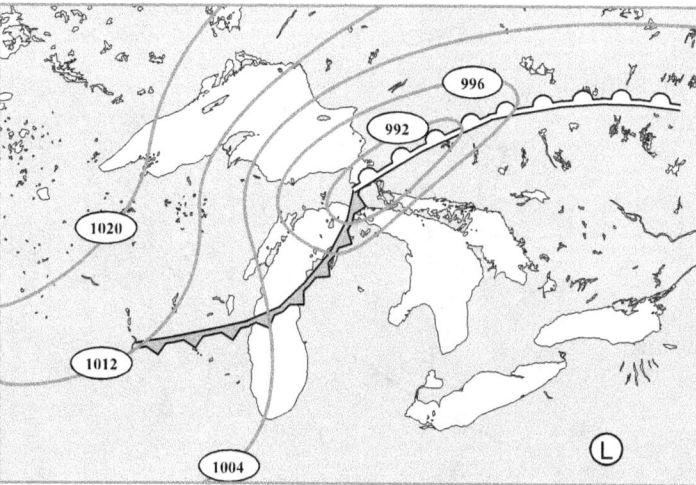

3. The advancing cold front follows the winds around the low pressure zone developing over Sault Ste Marie.

More warm air is force upward by both the advancing cold front and the slower warm front. The effect is greatest near the place where the fronts meet. More clouds form, more heating due to condensation. A low pressure cell is beginning to form. Rain is increasingly likely.

© Ross Lattner Publishing   www.rosslattner.ca

... on a Spinning Planet

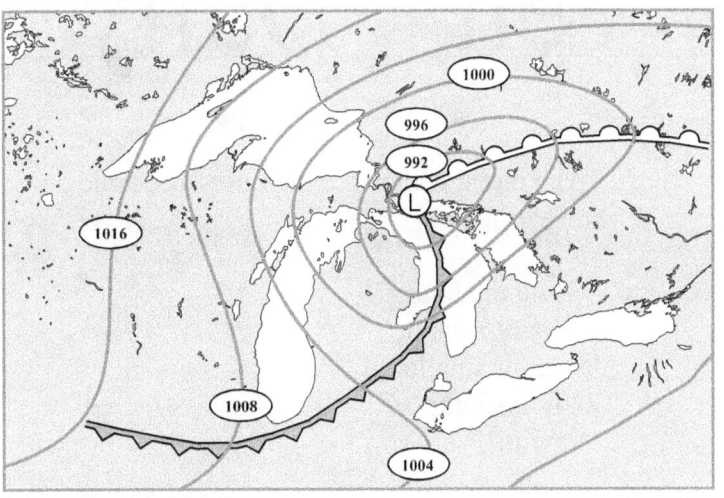

4 The rapidly advancing cold front is catching up with the slower warm front, creating a crotch. The low pressure cell is becoming quite well developed in the vicinity of the crotch.

The region within isobar 996 mb is quite cloudy. Rapid ascension of warm air parcels, followed by rapid condensation, is strengthening the warm, wet, low pressure cell

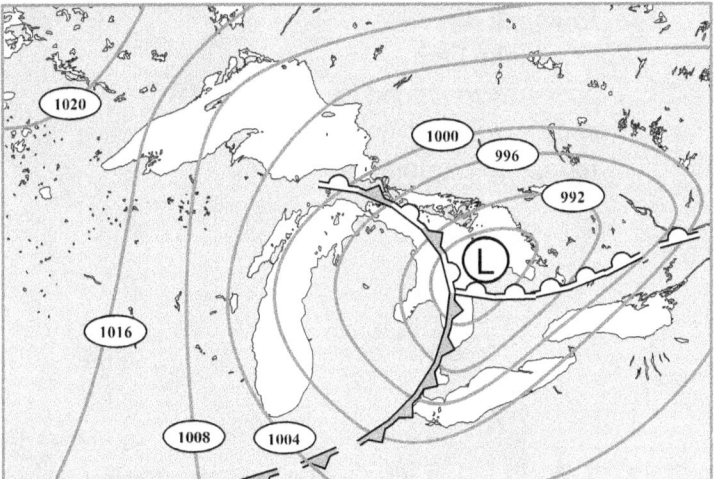

5 An "occluded front" begins to form. This accelerates the lofting of warm, moist air upward. Clouds and rain are almost certain over L. Huron and Georgian Bay.

Note that the original warm and cold fronts are beginning to disintegrate at their edges. More cold, dry air is moving in from the north.

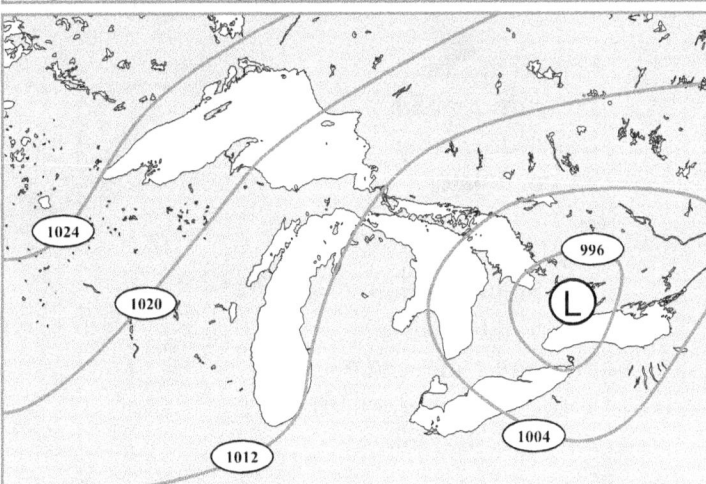

6 As the front system disintegrates, the low pressure system is pushed eastward, by cold, dry air moving in from the Northwest. The air parcels in the low pressure system become colder, dryer, and denser as the air rises. The remaining water falls in the Toronto area.. The low pressure cell becomes isolated, and gradually disappears.

Compare with diagram 1. Is it possible that the whole pattern is about to repeat itself?

For each of the diagrams above, draw arrows to indicate the wind direction along the isobars.

Use colored markers to indicate the regions of cloud formation. Indicate cumulus, cirrus and stratus clouds.

© Ross Lattner Publishing   113   www.rosslattner.ca

All the news that's fit to print... and then some

The Grade Ten Daily

Quiz 3.7: Spinning the Weather

1. The circle H marks a high pressure zone on the earth. The arrows indicate the speed of the spinning earth at each latitude.

   a. Is the high pressure circle expanding outward or is it contracting inward? Draw the high pressure zone as it might appear 24 hours later.

   b. In which direction would the high pressure zone tend to spin? Draw arrows to indicate. Explain.

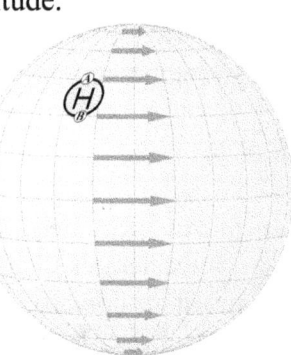

Date:

2. "The circle H marks a high pressure zone on the earth.

   a. Would the high pressure zone be larger or smaller 24 hours later? Draw it.

   b. In which direction would the high pressure zone tend to spin? Draw arrows to indicate. Explain.

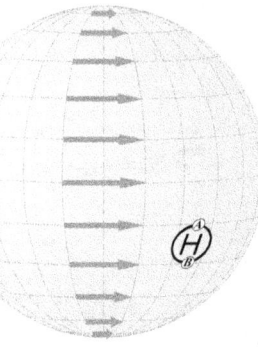

Date:

3. The circle L marks a low pressure zone.

   a. Would the low pressure zone be larger or smaller 24 hours later? Draw it.

   b. In which direction would the low pressure zone tend to spin? Draw arrows to indicate. Explain.

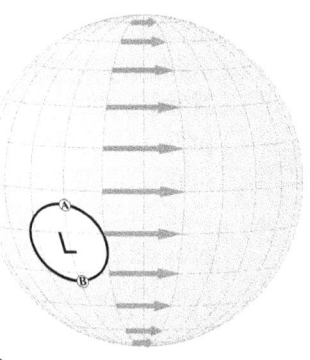

Date:

4. The circle L marks a low pressure zone.

   a. Would the low pressure zone be larger or smaller 24 hours later? Draw it.

   b. In which direction would the low pressure zone tend to spin? Draw arrows to indicate. Explain.

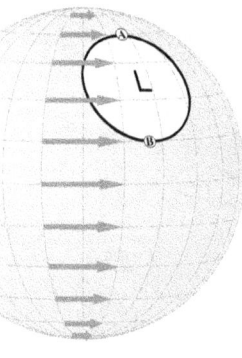

Date:

© Ross Lattner Publishing   www.rosslattner.ca

All the news that's fit to print... and then some

The Grade Ten Daily

Quiz 3.7: Spinning the Weather   Name:

Perfectly circular high pressure zones appear unlikely, don't you think? But the general principles remain the same. Consider the globes below, and draw arrows on the pressure lines to indicate the direction of circulation of air, due to the spinning earth.

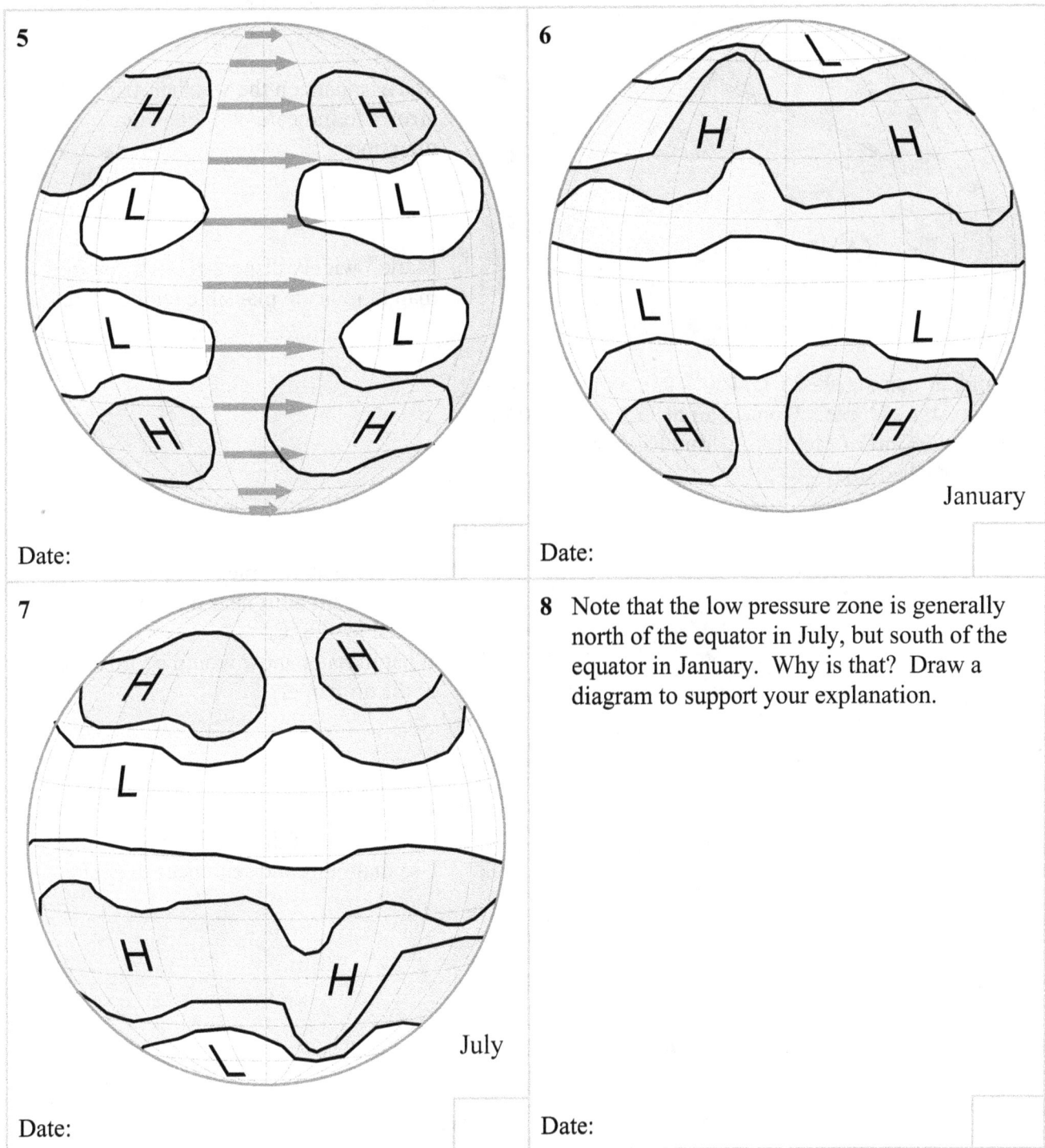

5.

6. January

7. July

8. Note that the low pressure zone is generally north of the equator in July, but south of the equator in January. Why is that? Draw a diagram to support your explanation.

Date: Date:

© Ross Lattner Publishing   www.rosslattner.ca

All the news that's fit to print... and then some

The Grade Ten Daily

Quiz 3.7: Spinning the Weather   Name:

It is late March. We have had three weeks of warm weather. Temperatures over souther Ontario average 22°C, and 60% RH. The air had been clear, but haze and clouds are beginning to obscure the sunshine. A mass of cold, dry air is moving in from the arctic circle.

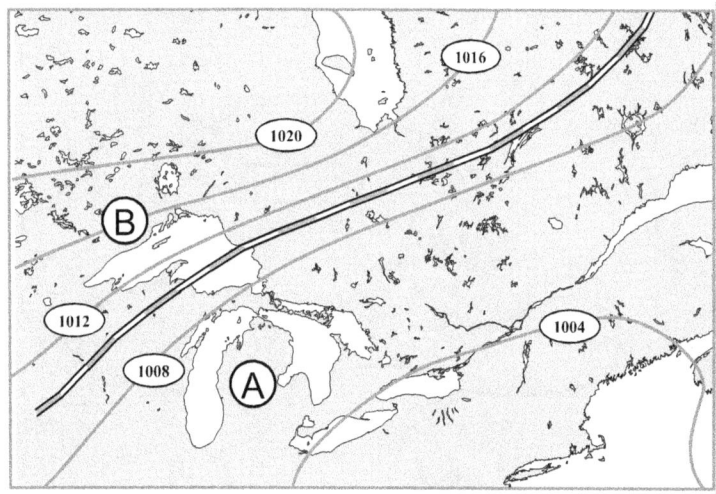

The cold air mass is at –25°C, 100% RH, and moving in from the Arctic Ocean. Tropical air at +22°C and 60% RH from the Gulf of Mexico. A front has begun to form along the dashed line.

9 What mass of water is contained in each tropical air parcel and in each arctic air parcel?

a. Mark isobars on the weather map with arrows to indicate possible wind directions.

b. Mark 5 widely dispersed points on the map to indicate plausible temperatures.

Date:

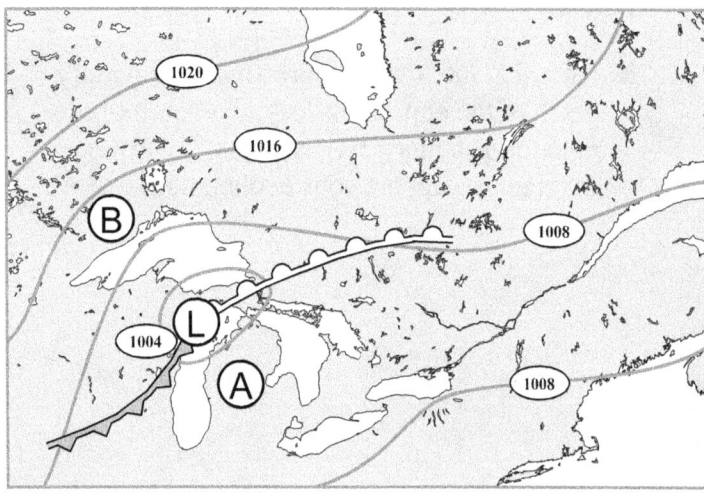

Twelve hours later, a front system has begun to form, with its center near Green Bay Wisconsin. The front system itself is drifting eastward along the great lakes, although the winds within it are blowing in various directions.

10 Suppose that a tropical air parcel is carried 5 km aloft near the front, and is cooled to –28°C.

a. What mass of snow would be frozen out of the air parcel?

b. Where, on the weather map, would you expect most of that snow to be falling? Use diagrams and sentences to explain.

Date:

© Ross Lattner Publishing   www.rosslattner.ca

All the news that's fit to print... and then some
The Grade Ten Daily

Quiz 3.7: Spinning the Weather   Name:

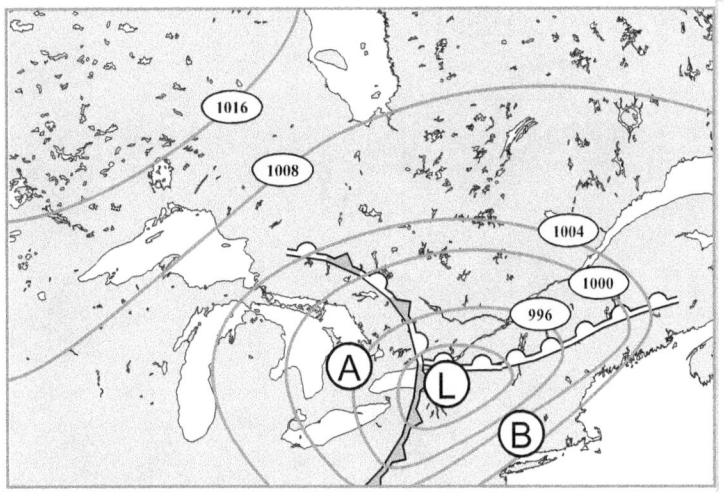

One day later, the tropical air has been pushed aloft to 5 km by the rapidly moving cold front. An occluded front has formed stretching from Toronto north to Superior.

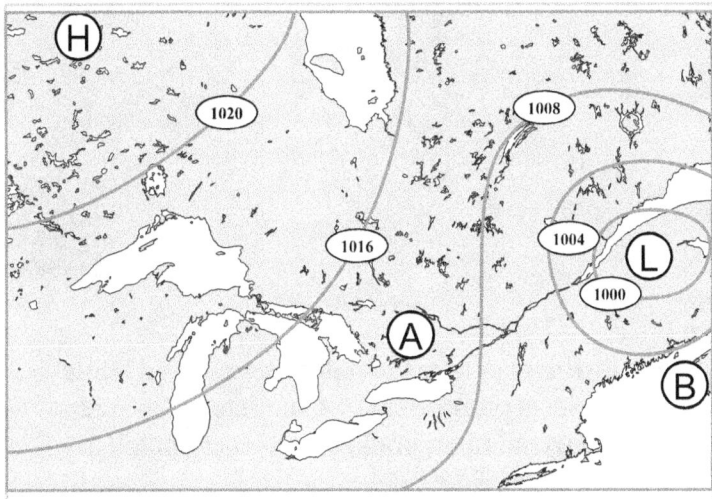

The low pressure cell has moved eastward along the Saint Lawrence. The energy that drove the front system has largely been spent. Cold dry air has moved in from the Arctic circle.

11 Draw arrows on the isobars to indicate the direction of the wind.

a. If 10 g of water vapour was to crystallize as snow, how much heat would be released in the 1.00 kg air parcel?

b. Calculate the ΔT of the air parcel, from the heat of solidification of the water vapour?

c. To what extent would this ΔT contribute to the intensity of any the low pressure cell?

Date:

12 Pick one location on the map. Describe the weather changes that have passed over that point in questions 9, 10, 11 and 12.

Date:

© Ross Lattner Publishing   117   www.rosslattner.ca

One step each day, done by Friday...
The Five Day Project

Project 4.1: Extreme Weather Events

0 **Project Instructions** Choose a weather event, and describe it. What is it like? Where and when is it found? Explain how that weather event "works." Where does the energy come from? (Include changes of state of water.) How do air (or water) parcels circulate and change in this weather event? How does the spinning of the earth provide direction for this kind of event?	0 My Plan and Outline Date:
1 In addition to your text, you must consult at least two books, two Internet sites, and one other resource of your choice. Prepare your library and internet search strategy. Execute your strategy, locating if possible more than you think you will need to use. Make proper bibliographic notes on each resource. Write a one page list of your sources.	1 Rough notes and plans here. Write enough that your teacher can understand what to expect in your one page report. Date:
2 Read your resources, and make notes. Compare and contrast resources, especially between newer and older sources. Organize your notes into a rough draft. Include discussion of the energy and circulation of the air parcels.	2 How do you plan to make notes as your read? File cards? Sheets of loose leaf? Computer? Be sure to include page number, book, etc. to every note!! Date:

One step each day, done by Friday...

The Five Day Project

Project 4.1: Extreme Weather Events   Name:

---

3 Write your report.

The whole report should be 600 – 1200 words, *plus* bibliography and cover.

In addition, you must include at least one full page picture, diagram, or photo of your subject.

---

3 Pay attention to organization. Make each paragraph count!

Date:

---

4 Give your report to someone else to read. At the same time, you read someone else's report.

Make comments upon spelling, grammar, clarity of writing, and so on.

Return the papers.

Polish up and revise your own report.

---

4 What do you need to fix in your report? Don't say "nothing". You can always improve.

Date:

---

5 Submit a finished report:

Cover. . . . . one page
Photo or picture. . . . . . . at least one page
Body of report. . . . . two to four pages
Bibliography . . . . one page

Attach this page for teacher assessment and evaluation.

---

5 After the project: What would you do differently next time?

Date:

Appendix: Laboratory Safety

10 Academic Science Lab Manual

The Hazards	The Safe Way
In this column is a list of lab safety issues that you will face in this course	Read this column to find out how to safely handle the laboratory problem.
Eye Injury is possible from flying fragments of metal, glass or chemicals; from heat or flames; from caustic solutions such as acids or bases.	*Always wear safety glasses* in the laboratory. Never take your glasses off, even if you have finished your experiment. Other students may not have finished theirs. The safety glass symbol indicates exercises in which safety glasses *must* be worn.
Crowding, Pushing and Horseplay increase the likelihood of a serious injury.	*Attend to your work.* Stay at the station you were assigned, so that there is room to work safely. If your teacher finds that your behaviour is a safety hazard, he or she may remove you from the lab. There is no place for behaviours which place others at risk of injury. Not at school, not at home and not at work.
Disorganized and Dirty Working Conditions are a hazard wherever they are found.	*Keep Lab Area Clean.* Clean and put away unused equipment. Tell your teacher about chipped, cracked, damaged or broken equipment. Do not leave anything on the floor, the desktop, the sink, or the cupboards that is not supposed to be there.
Broken Glass happens even to careful scientists.	*Do Not Touch* broken glass with your hands. Tell your teacher. Use a broom to sweep the glass into a dustpan. Dispose of the broken glass in the special container provided. Do not leave it in the regular wastebasket: it could seriously injure a custodian.
Liquid Spills may consist of water, but they may also contain acids, bases, or toxic chemicals. You may not be able to tell the difference.	*Tell your teacher* about any spills immediately. Do not attempt to clean up without teacher instruction. Only if the teacher decides it's safe, use a cloth or paper towels to soak up excess liquid. Wipe the area clean with a damp cloth. Rinse the cloth frequently in fresh water. Wash your hands afterwards.
Solid Spills may consist of highly reactive chemicals. You may not know the specific hazards.	*Tell Your Teacher* about the spill, whether or not you caused it. Your teacher will instruct you on the safe way to handle the problem. In any case, the spill must be cleaned up promptly.

Appendix: Laboratory Safety

Name:

Date:

Open Flames are a frequent hazard. The Bunsen burner is the most likely safety hazard.	***Review Safe Handling of Bunsen Burner*** with your teacher. Be prepared to show how to light, operate and extinguish the burner at any time. Do not attempt to ignite pens, papers, rulers or other things. That kind of behaviour will certainly result in your being put out of the lab.
Fire. Any liquid solid or gaseous fuel burning where you do not want it to burn is a fire.	***Tell the teacher immediately!*** Do not attempt to extinguish the fire with your hands, books, paper towels etc. Do not panic. Move away from the hazard. ***Your teacher is the best judge of the appropriate course of action.***
Hot Metal or Glass cause more burns than any other hazard. There is usually no visible indication that they are hot. Glass in particular causes small, deep burns.	***Let Hot Objects Cool for 10 - 15 Minutes*** before handling. Place all hot objects on a heat resistant pad. You and your partner will know where they are. Approach hot objects cautiously. Touch them at the coolest point first (the base of the retort rod, the bottom of the Bunsen burner or hot plate, the thumb screw of the iron ring). Use dry, not damp, paper towels to handle hot objects.
Hot Liquids such as boiling water or hot oil spread and splash rapidly. They also cling to skin and clothes.	***Let Hot Liquids Cool for 10 - 15 Minutes*** before handling. Do not heat liquids in closed containers. Use hot plates rather than shaky retort rod assemblies. Do not heat more liquid than you need.
Obstructed Passageways prevent you from moving out the way of a spill or a fire.	***Stand at Your Lab Station.*** Do not bring chairs or stools over to sit down. Your chair will prevent others from moving away from a spill or a fire.
Long Hair or Loose Clothing is more likely to become involved in your equipment. It can cause spills and breakage, or catch fire.	***Tie Back Long Hair; Secure Loose Clothing.*** Outerwear in particular must be avoided in the lab situation. Jackets, sweat suits, hoods, etc are too large and awkward for the lab situation. They are also frequently made of materials that are flammable and can melt and stick to the skin in a fire. Avoid using laquer based hair sprays. A curly head of hair with hair spray can burn up completely in seconds.
Unauthorized Experiments can have unintended results.	***Stick to the plan.*** Read instructions very carefully the night before the lab. Ask questions. Do not try experiments "just to see what happens." The dangers are too great.

www.ingramcontent.com/pod-product-compliance
Lightning Source LLC
Chambersburg PA
CBHW080444110426
42743CB00016B/3270